Brownian Brownian Motion-I

of the
American Mathematical Society

Number 927

Brownian Brownian Motion-I

N. Chernov
D. Dolgopyat

March 2009 • Volume 198 • Number 927 (fourth of 6 numbers) • ISSN 0065-9266

American Mathematical Society
Providence, Rhode Island

2000 *Mathematics Subject Classification.*
Primary 37D50; Secondary 34C29, 60F17.

Library of Congress Cataloging-in-Publication Data

Chernov, Nikolai, 1956–
  Brownian Brownian motion–I / Nikolai I. Chernov, Dmitry Dolgopyat.
    p. cm. — (Memoirs of the American Mathematical Society, ISSN 0065-9266 ; no. 927)
  "Volume 198, number 927 (fourth of 6 numbers)."
  Includes bibliographical references and index.
  ISBN 978-0-8218-4282-9 (alk. paper)
  1. Diffusion processes. 2. Brownian movements. 3. Limit theorems (Probability theory)
I. Dolgopyat, Dmitry, 1972– II. Title.
QA274.75.C456 2009
519.2′33—dc22   2008047662

Memoirs of the American Mathematical Society

This journal is devoted entirely to research in pure and applied mathematics.

Subscription information. The 2009 subscription begins with volume 197 and consists of six mailings, each containing one or more numbers. Subscription prices for 2009 are US$709 list, US$567 institutional member. A late charge of 10% of the subscription price will be imposed on orders received from nonmembers after January 1 of the subscription year. Subscribers outside the United States and India must pay a postage surcharge of US$65; subscribers in India must pay a postage surcharge of US$95. Expedited delivery to destinations in North America US$57; elsewhere US$160. Each number may be ordered separately; *please specify number* when ordering an individual number. For prices and titles of recently released numbers, see the New Publications sections of the *Notices of the American Mathematical Society*.

Back number information. For back issues see the *AMS Catalog of Publications*.

Subscriptions and orders should be addressed to the American Mathematical Society, P. O. Box 845904, Boston, MA 02284-5904, USA. *All orders must be accompanied by payment.* Other correspondence should be addressed to 201 Charles Street, Providence, RI 02904-2294, USA.

Copying and reprinting. Individual readers of this publication, and nonprofit libraries acting for them, are permitted to make fair use of the material, such as to copy a chapter for use in teaching or research. Permission is granted to quote brief passages from this publication in reviews, provided the customary acknowledgment of the source is given.

Republication, systematic copying, or multiple reproduction of any material in this publication is permitted only under license from the American Mathematical Society. Requests for such permission should be addressed to the Acquisitions Department, American Mathematical Society, 201 Charles Street, Providence, Rhode Island 02904-2294, USA. Requests can also be made by e-mail to reprint-permission@ams.org.

Memoirs of the American Mathematical Society (ISSN 0065-9266) is published bimonthly (each volume consisting usually of more than one number) by the American Mathematical Society at 201 Charles Street, Providence, RI 02904-2294, USA. Periodicals postage paid at Providence, RI. Postmaster: Send address changes to Memoirs, American Mathematical Society, 201 Charles Street, Providence, RI 02904-2294, USA.

© 2009 by the American Mathematical Society. All rights reserved.
Copyright of this publication reverts to the public domain 28 years
after publication. Contact the AMS for copyright status.
This publication is indexed in *Science Citation Index*®, *SciSearch*®, *Research Alert*®,
CompuMath Citation Index®, *Current Contents*®/*Physical, Chemical & Earth Sciences*.
Printed in the United States of America.

∞ The paper used in this book is acid-free and falls within the guidelines
established to ensure permanence and durability.
Visit the AMS home page at http://www.ams.org/

10 9 8 7 6 5 4 3 2 1   14 13 12 11 10 09

Contents

Chapter 1. Introduction	1
1.1. The model	1
1.2. The container	2
1.3. Billiard approximations	3
Chapter 2. Statement of results	7
2.1. Heavy disk in 'equilibrium' (linear motion)	7
2.2. Heavy disk at rest (slow acceleration)	9
2.3. Heavy disk of small size	11
2.4. Comparison to previous works	12
Chapter 3. Plan of the proofs	15
3.1. General strategy	15
3.2. Precise definitions	17
3.3. Key technical results	20
Chapter 4. Standard pairs and equidistribution	25
4.1. Unstable vectors	25
4.2. Unstable curves	29
4.3. Homogeneous unstable curves	33
4.4. Standard pairs	36
4.5. Perturbative analysis	42
4.6. Equidistribution properties	47
Chapter 5. Regularity of the diffusion matrix	57
5.1. Transport coefficients	57
5.2. Reduction to a finite series	62
5.3. Integral estimates: general scheme	64
5.4. Integration by parts	69
5.5. Cancellation of large boundary terms	77
5.6. Estimation of small boundary terms	81
5.7. Two-sided integral sums	84
5.8. Bounding off-diagonal terms	88
5.9. Hölder approximation	90
Chapter 6. Moment estimates	93

6.1.	General plan	93
6.2.	Structure of the proofs	98
6.3.	Short term moment estimates for V	100
6.4.	Moment estimates–*a priori* bounds	103
6.5.	Tightness	110
6.6.	Second moment	113
6.7.	Martingale property	115
6.8.	Transition to continuous time	117
6.9.	Uniqueness for stochastic differential equations	118

Chapter 7. Fast slow particle — 123

Chapter 8. Small large particle — 129

Chapter 9. Open problems — 133
- 9.1. Collisions of the massive disk with the wall — 133
- 9.2. Longer time scales — 133
- 9.3. Stadia and the piston problem — 133
- 9.4. Finitely many particles — 134
- 9.5. Growing number of particles — 135
- 9.6. Particles of positive size — 136

Appendix A. Statistical properties of dispersing billiards — 139
- A.1. Decay of correlations: overview — 139
- A.2. Decay of correlations: extensions — 149
- A.3. Large deviations — 153
- A.4. Moderate deviations — 154
- A.5. Nonsingularity of diffusion matrix — 158
- A.6. Asymptotics of diffusion matrix — 159

Appendix B. Growth and distortion in dispersing billiards — 167
- B.1. Regularity of H-curves — 167
- B.2. Invariant Section Theorem — 171
- B.3. The function space $\mathfrak{R}$ — 174

Appendix C. Distortion bounds for two particle system — 177

Bibliography — 187

Index — 193

Abstract

A classical model of Brownian motion consists of a heavy molecule submerged into a gas of light atoms in a closed container. In this work we study a 2D version of this model, where the molecule is a heavy disk of mass $M \gg 1$ and the gas is represented by just one point particle of mass $m = 1$, which interacts with the disk and the walls of the container via elastic collisions. Chaotic behavior of the particles is ensured by convex (scattering) walls of the container. We prove that the position and velocity of the disk, in an appropriate time scale, converge, as $M \to \infty$, to a Brownian motion (possibly, inhomogeneous); the scaling regime and the structure of the limit process depend on the initial conditions. Our proofs are based on strong hyperbolicity of the underlying dynamics, fast decay of correlations in systems with elastic collisions (billiards), and methods of averaging theory.

---

[1]Received by editor May 15, 2005; and in revised form September 18, 2006.
[2]Key words: dispersing billiards, averaging, shadowing, diffusion processes.
[3]2000 Mathematics Subject Classification: 37D50; 34C29, 60F17.
[4]Nikolai Chernov: Department of Mathematics, University of Alabama at Birmingham, Birmingham, AL 35294.
[5]Dmitry Dolgopyat: Department of Mathematics, University of Maryland, College Park 20742.

CHAPTER 1

Introduction

1.1. The model. We study a dynamical system of two particles – a hard disk of radius $\mathbf{r} > 0$ and mass $M \gg 1$ and a point particle of mass $m = 1$. Our particles move freely in a two-dimensional container $\mathcal{D}$ with concave boundaries and collide elastically with each other and with the walls (boundary) of $\mathcal{D}$. Our assumptions on the shape of the container $\mathcal{D}$ are stated in Section 1.2.

Let $Q(t)$ denote the center and $V(t)$ the velocity of the heavy disk at time t. Similarly, let $q(t)$ denote the position of the light particle and $v(t)$ its velocity. When a particle collides with a scatterer, the normal component of its velocity reverses. When the two particles collide with each other, the normal components of their velocities change by the rules

$$(1.1) \qquad v_{\text{new}}^{\perp} = -\frac{M-1}{M+1} v_{\text{old}}^{\perp} + \frac{2M}{M+1} V_{\text{old}}^{\perp}$$

and

$$(1.2) \qquad V_{\text{new}}^{\perp} = \frac{M-1}{M+1} V_{\text{old}}^{\perp} + \frac{2}{M+1} v_{\text{old}}^{\perp},$$

while the tangential components remain unchanged. The total kinetic energy is conserved, and we fix it so that

$$(1.3) \qquad \|v\|^2 + M\|V\|^2 = 1.$$

This implies $\|v\| \leq 1$ and $\|V\| \leq 1/\sqrt{M}$.

This is a Hamiltonian system, and it preserves Liouville measure on its phase space. Systems of hard disks in closed containers are proven to be completely hyperbolic and ergodic under various conditions [**12, 81, 82, 83**]. These results do not cover our particular model, but we have little doubt that it is hyperbolic and ergodic, too. In this paper, though, we do not study ergodic properties.

We are interested in the evolution of the system during the initial period of time before the heavy disk experiences its first collision with the border $\partial \mathcal{D}$. This condition restricts our analysis to an interval of time $(0, cM^a)$, where $c, a > 0$ depend on $Q(0)$ and $V(0)$, see Chapter 2.

During this initial period, the system does not exhibit its ergodic behavior, but it does exhibit a diffusive behavior in the following sense. As (1.1)–(1.2) imply,

$$(1.4) \qquad \left|\|v_{\text{new}}\| - \|v_{\text{old}}\|\right| \leq 2/\sqrt{M} \quad \text{and} \quad \|V_{\text{new}} - V_{\text{old}}\| \leq 2/M,$$

hence the changes in $\|v\|$ and V at each collision are much smaller than their typical values, which are $\|v\| = \mathcal{O}(1)$ and $\|V\| = \mathcal{O}(1/\sqrt{M})$. Thus, the speed of the light particle, $\|v(t)\|$, remains almost constant, and the heavy particle not only moves slowly but its velocity $V(t)$ changes slowly as well (it has *inertia*). We will show that, in the limit $M \to \infty$, the velocity $V(t)$ can be approximated by a Brownian motion, and the position $Q(t)$ by an integral of the Brownian motion.

Our system is one of the simplest models of a particle moving in a fluid. This is what scientists called Brownian motion about one hundred years ago. Now this term has a more narrow technical meaning, namely a Gaussian process with zero mean and stationary independent increments. Our paper is motivated by the Brownian motion in its original sense, and this is why we call it "Brownian Brownian motion".

Even though this paper only covers a two particle system (where the "fluid" is represented by a single light particle), we believe that our methods can extend to more realistic fluids of many particles. We consider this paper as a first step in our studies (thus the numeral one in its title) and plan to investigate more complex models in the future, see our discussion of open problems in Chapter 9.

1.2. The container. In this paper we assume that $\mathcal{D}$ is a dispersing billiard table with finite horizon and smooth boundary:

Assumption A1: $\mathcal{D}$ is a dispersing billiard table, i.e. its boundary $\partial \mathcal{D}$ is concave; this guarantees chaotic motion of the particle colliding with $\partial \mathcal{D}$.

Assumption A2: $\mathcal{D}$ has finite horizon, which means the point particle cannot travel longer than a certain finite distance $L_{\max} < \infty$ without collisions (even if we remove the hard disk from $\mathcal{D}$); this prevents superdiffusive (ballistic) motion of the particle [4].

Assumption A3: $\mathcal{D}$ has C^3 smooth boundary (without corner points).

Containers satisfying all these assumptions can be constructed as follows. Let $\mathbb{T}^2$ be the unit torus and $\mathbb{B}_1, \ldots, \mathbb{B}_r \subset \mathbb{T}^2$ some disjoint convex regions with C^3 smooth boundaries, whose curvature never vanishes. Then

$$\mathcal{D} = \mathbb{T}^2 \setminus \cup_{i=1}^r \mathbb{B}_i.$$

The obstacles $\mathbb{B}_1,\ldots,\mathbb{B}_r$ act as scatterers, our light particle bounces between them (like in a pinball machine). They must also block all collision-free flights of the particle to ensure the finite horizon assumption.

The motion of a single particle in such domains $\mathcal{D}$ has been studied by Ya. Sinai [86], and this model is now known as dispersing billiard system. It is always hyperbolic and ergodic [86], and has strong statistical properties [11, 96].

Our assumptions on $\mathcal{D}$ are fairly restrictive. It would be tempting to cover simpler containers – just a rectangular box, for example. We believe that most of our results would carry over to rectangular boxes (perhaps, with certain adjustments). However, a two-particle system in a rectangular container, despite its apparent simplicity, would be much more difficult to analyze, because the corresponding billiard system is not chaotic. For this reason rectangular containers are currently out of reach. On the other hand if the boundary of $\mathcal{D}$ is convex then with positive probability the particles will never meet (see [64]), so some assumptions on the shape of $\mathcal{D}$ are necessary.

1.3. Billiard approximations. We denote phase points by $x = (Q, V, q, v)$ and the phase space by $\mathcal{M}$. Due to the energy conservation (1.3), $\dim \mathcal{M} = 7$. The dynamics $\Phi^t\colon \mathcal{M} \to \mathcal{M}$ can be reduced, in a standard way, to a discrete time system – a collision map – as follows.

We call $\Omega = \partial\mathcal{M}$ the *collision space*. Let $\mathcal{P}(Q)$ denote the disk of radius $\mathbf{r}$ centered on Q, then $\Omega = \{(Q, V, q, v) \in \mathcal{M}\colon q \in \partial\mathcal{D} \cup \partial\mathcal{P}(Q)\}$. At each collision, we identify the precollisional and postcollisional velocity vectors. Technically, we will only include the *postcollisional* vector in Ω, so that

$$\Omega = \Omega_\mathcal{D} \cup \Omega_\mathcal{P},$$
$$\Omega_\mathcal{D} = \{(Q, V, q, v) \in \mathcal{M}\colon q \in \partial\mathcal{D},\ \langle v, n\rangle \geq 0\},$$
$$\Omega_\mathcal{P} = \{(Q, V, q, v) \in \mathcal{M}\colon q \in \partial\mathcal{P}(Q),\ \langle v - V, n\rangle \geq 0\}.$$

where $\langle \cdot, \cdot \rangle$ stands for the scalar product of vectors and n denotes a normal vector to $\partial\mathcal{D} \cup \partial\mathcal{P}(Q)$ at q pointing into $\mathcal{D} \setminus \mathcal{P}(Q)$. The first return map $\mathcal{F}\colon \Omega \to \Omega$ is called the *collision map*. It preserves a smooth probability measure μ on Ω induced by the Liouville measure on $\mathcal{M}$.

It will be convenient to denote points of Ω by (Q, V, q, w), where

$$(1.5) \qquad w = \begin{cases} v & \text{for } q \in \partial\mathcal{D} \\ v - V & \text{for } q \in \partial\mathcal{P}(Q) \end{cases}$$

so that Ω can be represented by a unified formula

$$
(1.6) \qquad \Omega = \{(Q, V, q, w) \colon q \in \partial \mathcal{D} \cup \partial \mathcal{P}(Q),\ \langle w, n \rangle \geq 0\}
$$

For every point $(Q, V, q, w) \in \Omega$ we put

$$
(1.7) \qquad \bar{w} = \frac{w}{\|w\|}\, \mathbf{s}_V, \qquad \mathbf{s}_V = \sqrt{1 - M\|V\|^2}
$$

(note that $\|\bar{w}\| = \mathbf{s}_V = \|v\|$ due to (1.3), hence $\mathbf{s}_V$ only depends on $\|V\|$). For each pair (Q, V) we denote by $\Omega_{Q,V}$ the cross-section of Ω obtained by fixing Q and V. By using (1.7) we can write

$$
(1.8) \qquad \Omega_{Q,V} = \{(q, \bar{w}) \colon q \in \partial \mathcal{D} \cup \partial \mathcal{P}(Q),\ \langle \bar{w}, n \rangle \geq 0,\ \|\bar{w}\| = \mathbf{s}_V\}
$$

Now let us pick $t_0 \geq 0$ and fix the center of the heavy disk at $Q = Q(t_0) \in \mathcal{D}$ and set $M = \infty$. Then the light particle would move with a constant speed $\|v(t)\| = \mathbf{s}_V$, where $V = V(t_0)$, in the domain $\mathcal{D} \setminus \mathcal{P}(Q)$ with specular reflections at $\partial \mathcal{D} \cup \partial \mathcal{P}(Q)$. Thus we get a billiard-type dynamics, which approximates our system during a relatively short interval of time, until our heavy disk moves a considerable distance. We may treat our system then as a small perturbation of this billiard-type dynamics, and in fact our entire analysis is based on this approximation.

The collision map $\mathcal{F}_{Q,V}$ of the above billiard system acts on the space (1.8), where $\bar{w}$ denotes the postcollisional velocity of the moving particle. The map $\mathcal{F}_{Q,V} \colon \Omega_{Q,V} \to \Omega_{Q,V}$ preserves a smooth probability measure $\mu_{Q,V}$ as described in Chapter 4.

The billiard-type system $(\Omega_{Q,V}, \mathcal{F}_{Q,V}, \mu_{Q,V})$ is essentially independent of V. By a simple rescaling (i.e. renormalizing) of $\bar{w}$ we can identify it with $(\Omega_{Q,0}, \mathcal{F}_{Q,0}, \mu_{Q,0})$, which we denote, for brevity, by $(\Omega_Q, \mathcal{F}_Q, \mu_Q)$, and it becomes a standard billiard system, where the particle moves at unit speed, on the table $\mathcal{D} \setminus \partial \mathcal{P}(Q)$. This is a dispersing (Sinai) table, hence the map $\mathcal{F}_Q$ is hyperbolic, ergodic and has strong statistical properties, including exponential decay of correlations and the central limit theorem [**86, 96**].

Equations (1.1)–(1.2) imply that the change of the velocity of the disk due to a collision with the light particle is

$$
(1.9) \qquad V_{\text{new}} - V_{\text{old}} = -\frac{2\left(v_{\text{new}}^{\perp} - V_{\text{new}}^{\perp}\right)}{M+1} = -\frac{2\, w^{\perp}}{M+1}
$$

Since $w = \bar{w}\, \|v - V\|/\|v\|$, we have

$$
(1.10) \qquad V_{\text{new}} - V_{\text{old}} = -\frac{2\, \bar{w}^{\perp}}{M} + \delta
$$

where

$$|\delta| \leq \text{Const}\left(\frac{\|V\|}{M\|v\|} + \frac{1}{M^2}\right) \tag{1.11}$$

is a relatively small term. Define a vector function on Ω by

$$\mathcal{A} = \begin{cases} -2\bar{w}^\perp & \text{for } q \in \partial \mathcal{P}(Q) \\ 0 & \text{for } q \in \partial \mathcal{D} \setminus \partial \mathcal{P}(Q) \end{cases} \tag{1.12}$$

Obviously, $\mathcal{A}$ is a smooth function, and due to a rotational symmetry

$$\int \mathcal{A} \, d\mu_{Q,V} = 0$$

for every Q, V. Hence the central limit theorem for dispersing billiards [9, 96] implies the convergence in distribution

$$\frac{1}{\sqrt{n}} \sum_{j=0}^{n-1} \mathcal{A} \circ \mathcal{F}_{Q,V}^j \to \mathcal{N}(0, \bar{\sigma}_{Q,V}^2(\mathcal{A})). \tag{1.13}$$

as $n \to \infty$, where $\bar{\sigma}_{Q,V}^2(\mathcal{A})$ a symmetric positive semidefinite matrix given by the Green-Kubo formula

$$\bar{\sigma}_{Q,V}^2(\mathcal{A}) = \sum_{j=-\infty}^{\infty} \int_{\Omega_{Q,V}} \mathcal{A} \left(\mathcal{A} \circ \mathcal{F}_{Q,V}^j\right)^T d\mu_{Q,V}. \tag{1.14}$$

(this series converges because its terms decay exponentially fast as $|j| \to \infty$, see [96, 18]). By setting $V = 0$ we define a matrix $\bar{\sigma}_Q^2(\mathcal{A}) := \bar{\sigma}_{Q,0}^2(\mathcal{A})$. Since the restriction of $\mathcal{A}$ to the sets $\Omega_{Q,V}$ and $\Omega_{Q,0} = \Omega_Q$ only differ by a scaling factor s_V, see (1.7), we have a simple relation

$$\bar{\sigma}_{Q,V}^2(\mathcal{A}) = (1 - M\|V\|^2) \, \bar{\sigma}_Q^2(\mathcal{A}). \tag{1.15}$$

Define another matrix

$$\sigma_Q^2(\mathcal{A}) = \bar{\sigma}_Q^2(\mathcal{A})/\bar{L}, \tag{1.16}$$

where

$$\bar{L} = \pi \frac{\text{Area}(\mathcal{D}) - \text{Area}(\mathcal{P})}{\text{length}(\partial \mathcal{D}) + \text{length}(\partial \mathcal{P})} \tag{1.17}$$

is the mean free path of the light particle in the billiard dynamics $\mathcal{F}_Q$, see [16] (observe that $\bar{L}$ does not depend on Q). Lastly, let $\sigma_Q(\mathcal{A})$ be the symmetric positive semidefinite square root of $\sigma_Q^2(\mathcal{A})$.

Let us draw some conclusions, which will be entirely heuristic at this point (they will be formalized later). In view of (1.9)–(1.13), we may expect that the total change of the disk velocity V in the course of n consecutive collisions of the light particle with $\partial \mathcal{D} \cup \partial \mathcal{P}$ can be approximated by a normal random variable $\mathcal{N}\left(0, n\bar{\sigma}_{Q,V}^2(\mathcal{A})/M^2\right)$. During an interval (t_0, t_1), the light particle experiences $n \approx \bar{L}^{-1}\|v\|(t_1 - t_0)$ collisions, hence for the total change of the disk velocity we expect another normal approximation

$$(1.18) \qquad V(t_1) - V(t_0) \sim \mathcal{N}\left(0, \|v\|(t_1 - t_0)\, \sigma_{Q,V}^2(\mathcal{A})/M^2\right)$$

(due to the inertia of the heavy disk, we expect $Q(t) \approx Q(t_0)$, $\|V(t)\| \approx \|V(t_0)\|$, and thus $\|v(t)\| \approx \mathbf{s}_V(t_0)$ for all $t_0 < t < t_1$). A large part of our paper is devoted to making the heuristic approximation (1.18) precise.

CHAPTER 2

Statement of results

Suppose the initial position $Q(0) = Q_0$ and velocity $V(0) = V_0$ of the heavy particle are fixed, and the initial state of the light particle $q(0), v(0)$ is selected randomly, according to a smooth distribution in the direct product of the domain $\mathcal{D} \setminus \mathcal{P}(Q_0)$ and the circle $\|v(0)\|^2 = 1 - M\|V_0\|^2$ (alternatively, $q(0)$ may be chosen from $\partial \mathcal{D} \cup \partial \mathcal{P}(Q_0)$ and $v(0)$ from the semicircle containing all the postcollisional velocity vectors). The shape of the initial distribution will not affect our results.

We consider the trajectory of the heavy particle $Q(t), V(t)$ during a time interval $(0, cM^a)$ with some $c, a > 0$ selected below. We scale time by $\tau = t/M^a$ and, sometimes, scale space in a way specified below, to convert $\{Q(t), V(t)\}$ to a pair of functions $\{\mathcal{Q}(\tau), \mathcal{V}(\tau)\}$ on the interval $0 < \tau < c$. The random choice of $q(0), v(0)$ induces a probability measure on the space of functions $\mathcal{Q}(\tau), \mathcal{V}(\tau)$, and we are interested in the convergence of this probability measure, as $M \to \infty$, to a stochastic process $\{\mathbf{Q}(\tau), \mathbf{V}(\tau)\}$. We prove three major results in this direction corresponding to three different regimes in the dynamics of the massive disk.

2.1. Heavy disk in 'equilibrium' (linear motion). First, let the initial velocity of the heavy particle be of order $1/\sqrt{M}$. Specifically, let us fix a unit vector $u_0 \in S^1$ and $\chi \in (0, 1)$, set

(2.1) $$V_0 = M^{-1/2} \chi u_0$$

and fix $Q_0 \in \mathcal{D}$ arbitrarily (but so that $\text{dist}(Q_0, \partial \mathcal{D}) > \mathbf{r}$). Note that if the heavy disk moved with a constant velocity, V_0, without colliding with the light particle, it would hit $\partial \mathcal{D}$ at a certain moment $c_0 M^{1/2}$, where $c_0 > 0$ is determined by Q_0, u_0 and χ. We restrict our analysis to a time interval $(0, cM^{1/2})$ with some $c < c_0$. During this period of time we expect, due to (1.18), that the overall fluctuations of the disk velocity will be $\mathcal{O}(M^{-3/4})$. Hence we expect $V(t) = V_0 + \mathcal{O}(M^{-3/4})$ and $Q(t) = Q_0 + tV_0 + \mathcal{O}(tM^{-3/4}) = Q_0 + tV_0 + \mathcal{O}(M^{-1/4})$ for $0 < t < cM^{1/2}$. This leads us to a time scale

(2.2) $$\tau = tM^{-1/2}$$

and a space scale

$$\mathcal{Q}(\tau) = M^{1/4}\left[Q(\tau M^{1/2}) - Q_0 - \tau M^{1/2}V_0\right] \tag{2.3}$$

and, respectively,

$$\mathcal{V}(\tau) = M^{3/4}\left[V(\tau M^{1/2}) - V_0\right] \tag{2.4}$$

We can find an asymptotic distribution of $\mathcal{V}(\tau)$ by using our heuristic normal approximation (1.18). Let

$$Q^\dagger(\tau) = Q_0 + \tau M^{1/2}V_0 = Q_0 + \tau\chi\, u_0$$

Then for any $\tau \in (0, c)$ we have $Q(\tau M^{1/2}) \to Q^\dagger(\tau)$, as $M \to \infty$, hence

$$\sigma^2_{Q(\tau M^{1/2})}(\mathcal{A}) \to \sigma^2_{Q^\dagger(\tau)}(\mathcal{A}). \tag{2.5}$$

Anticipating that $\|v(t)\| \approx \sqrt{1-\chi^2}$ for all $0 < t < cM^{1/2}$ we can expect that for small $d\tau > 0$ the number of collisions $N(d\tau)$ is approximately equal to $d\tau \bar{L}^{-1}\sqrt{1-\chi^2}$ and the momenta exchange during each collision is close to $\sqrt{1-\chi^2}\mathcal{A}$, see (1.12). This should give us

$$\mathcal{V}(\tau + d\tau) - \mathcal{V}(\tau) \sim \mathcal{N}\bigl(0, D(d\tau)\bigr)$$

where

$$D(d\tau) \approx N(d\tau)\,(1-\chi^2)\,\bar{\sigma}^2_{Q^\dagger(\tau)}(\mathcal{A})$$
$$= d\tau\,\sqrt{(1-\chi^2)^3}\,\sigma^2_{Q^\dagger(\tau)}(\mathcal{A}) \tag{2.6}$$

Integrating (2.6) over $(0, \tau)$ yields

$$\mathcal{V}(\tau) \sim \mathcal{N}\left(0, \sqrt{(1-\chi^2)^3}\int_0^\tau \sigma^2_{Q^\dagger(s)}(\mathcal{A})\,ds\right) \tag{2.7}$$

The following theorem (proved in this paper) makes this conclusion precise:

THEOREM 1. *Under the conditions A1–A3 and (2.1) the random process $\mathcal{V}(\tau)$ defined on the interval $0 \leq \tau \leq c$ by (2.2), (2.4) weakly converges, as $M \to \infty$, to a Gaussian Markov random process $\mathbf{V}(\tau)$ with independent increments, zero mean and covariance matrix*

$$\mathrm{Cov}(\mathbf{V}(\tau)) = \sqrt{(1-\chi^2)^3}\int_0^\tau \sigma^2_{Q^\dagger(s)}(\mathcal{A})\,ds \tag{2.8}$$

The process $\mathbf{V}(\tau)$ can be, equivalently, defined by

$$\mathbf{V}(\tau) = \sqrt[4]{(1-\chi^2)^3}\int_0^\tau \sigma_{Q^\dagger(s)}(\mathcal{A})\,d\mathbf{w}(s) \tag{2.9}$$

where $\mathbf{w}(s)$ denotes the standard two dimensional Brownian motion. Accordingly, the function $\mathcal{Q}(\tau)$ defined by (2.3) converges weakly to a

Gaussian random process $\mathbf{Q}(\tau) = \int_0^\tau \mathbf{V}(s)\,ds$, which has zero mean and covariance matrix

$$(2.10) \qquad \mathrm{Cov}(\mathbf{Q}(\tau)) = \sqrt{(1-\chi^2)^3} \int_0^\tau (\tau-s)^2 \sigma^2_{Q^\dagger(s)}(\mathcal{A})\,ds$$

We note that the limit velocity process is a (time inhomogeneous) Brownian motion, while the limit position process is the integral of that Brownian motion.

2.2. Heavy disk at rest (slow acceleration).

The next theorem deals with the more difficult case of

$$(2.11) \qquad V_0 = 0$$

(here again, Q_0 is chosen arbitrarily). Since the heavy disk is now initially at rest, it takes it longer to build up speed and travel to the border $\partial\mathcal{D}$. We expect, due to (1.18), that the disk velocity grows as $\|V(t)\| = \mathcal{O}(\sqrt{t}/M)$, and therefore its displacement grows as $\|Q(t) - Q_0\| = \mathcal{O}(t^{3/2}/M)$. Hence, for typical trajectories, it takes $\mathcal{O}(M^{2/3})$ units of time for the disk to reach $\partial\mathcal{D}$. It is convenient to modify the dynamics of the disk making it stop ("freeze") when it comes too close to $\partial\mathcal{D}$. We pick a small $\delta_0 > 0$ and stop the disk at the moment

$$(2.12) \qquad t_* = \min\{t > 0\colon \mathrm{dist}(Q(t), \partial\mathcal{D}) = \mathbf{r} + \delta_0\},$$

hence we obtain a modified dynamics $Q_*(t)$, $V_*(t)$ given by

$$Q_*(t) = \begin{cases} Q(t) & \text{for } t < t_* \\ Q(t_*) & \text{for } t > t_* \end{cases} \qquad V_*(t) = \begin{cases} V(t) & \text{for } t < t_* \\ 0 & \text{for } t > t_* \end{cases}$$

With this modification, we can consider the dynamics on a time interval $(0, cM^{2/3})$ with an arbitrary $c > 0$. Our time scaling is

$$(2.13) \qquad \tau = tM^{-2/3}$$

and there is no need for any space scaling, i.e. we set

$$(2.14) \qquad \mathcal{Q}(\tau) = Q_*(\tau M^{2/3}) \qquad \text{and} \qquad \mathcal{V}(\tau) = M^{2/3} V_*(\tau M^{2/3}).$$

We note that the heavy disk, being initially at rest, can move randomly in any direction and follow a random trajectory before reaching $\partial\mathcal{D}$. Thus, the matrix $\sigma^2_{Q(\tau M^{2/3})}(\mathcal{A})$ does not have a limit (in any sense), as $M \to \infty$, because it depends on the (random) location of the disk. Hence, heuristic estimates of the sort (2.5)–(2.7) are now impossible, and the limit distribution of the functions $\{\mathcal{Q}(\tau), \mathcal{V}(\tau)\}$ cannot be found explicitly. Instead, we will show that any weak limit of these functions, call it $\{\mathbf{Q}(\tau), \mathbf{V}(\tau)\}$, satisfies two stochastic differential equations (SDE)

$$(2.15) \qquad d\mathbf{Q} = \mathbf{V}\, d\tau, \qquad d\mathbf{V} = \sigma_{\mathbf{Q}}(\mathcal{A})\, d\mathbf{w}(\tau)$$

with initial conditions $\mathbf{Q}_0 = Q_0$ and $\mathbf{V}_0 = 0$. Thus, the limit behavior of the functions $\mathcal{Q}(\tau)$ and $\mathcal{V}(\tau)$ will only be described implicitly, via (2.15).

In order to guarantee the convergence, though, we need to make sure that the initial value problem (2.15) has a unique solution $\{\mathbf{Q}(\tau), \mathbf{V}(\tau)\}$. SDE of this type have unique solutions if the matrix $\sigma_Q(\mathcal{A})$, as a function of Q, is differentiable [**76**, Section IX.2] but they may have multiple solutions if σ_Q is only continuous. We do not expect our matrix $\sigma_Q(\mathcal{A})$ to be differentiable, though. Recent numerical experiments [**5**] suggest that dynamical invariants, such as the diffusion matrix, are not differentiable if the system has singularities.

To tackle this problem, we prove (see Section 6.9) that the SDE (2.15) has a unique solution provided $\sigma_Q(\mathcal{A})$ is log-Lipschitz continuous in the following sense:

$$(2.16) \qquad \|\sigma_{Q_1}(\mathcal{A}) - \sigma_{Q_2}(\mathcal{A})\| \leq \mathrm{Const}\, \|Q_1 - Q_2\| \left| \ln \|Q_1 - Q_2\| \right|$$

(this condition is weaker than Lipschitz continuity but stronger than Hölder continuity with any exponent < 1).

Thus we need to establish (2.16), which constitutes a novel and rather difficult result in billiard theory. Its proof occupies a sizable part of our paper (Chapter 5) and requires two additional assumptions on the scatterers $\mathbb{B}_i$. First, the collision map $\mathcal{F}_Q \colon \Omega_Q \to \Omega_Q$ must be C^3 smooth (rather than C^2, which is commonly assumed in the studies of billiards), hence the boundaries $\partial \mathbb{B}_i$ must be at least C^4. This additional smoothness of $\mathcal{F}_Q$ allows us to prove that

$$(2.17) \qquad \|\sigma^2_{Q_1}(\mathcal{A}) - \sigma^2_{Q_2}(\mathcal{A})\| \leq \mathrm{Const}\, \|Q_1 - Q_2\| \left| \ln \|Q_1 - Q_2\| \right|$$

which is slightly weaker than (2.16). To convert (2.17) to (2.16) we need the matrix $\sigma^2_Q(\mathcal{A})$ be nonsingular for every Q, so that the function $\sigma^2 \mapsto \sigma$ is smooth. To this end we only find a criterion (see Section A.5), in terms of periodic orbits of $\mathcal{F}_Q$, for the nonsingularity of $\sigma^2_Q(\mathcal{A})$. We believe it is satisfied for typical configurations of scatterers $\mathbb{B}_i$, but we do not prove it here.

Thus we formulate our additional assumptions:

Assumption A3': The boundaries $\partial \mathbb{B}_i$ of all scatterers are C^4 smooth;

Assumption A4: $\sigma^2_Q(\mathcal{A}) > 0$ for all $Q \in \mathcal{D}$ such that $\mathrm{dist}(\mathcal{P}(Q), \partial\mathcal{D}) \geq \delta_0$.

Next we state the convergence theorem:

THEOREM 2. *Under the conditions A1, A2, A3', A4 the random processes $\{\mathcal{Q}(\tau), \mathcal{V}(\tau)\}$ defined on the interval $0 \leq \tau \leq c$ by (2.11)–(2.14) weakly converge to a stochastic process $\{\mathbf{Q}(\tau), \mathbf{V}(\tau)\}$, which constitutes a unique solution of the following stochastic differential equations with initial conditions:*

$$(2.18) \qquad \begin{aligned} d\mathbf{Q} &= \mathbf{V}\, d\tau, & \mathbf{Q}_0 &= Q_0 \\ d\mathbf{V} &= \sigma_{\mathbf{Q}}(\mathcal{A})\, d\mathbf{w}(\tau), & \mathbf{V}_0 &= 0 \end{aligned}$$

which are stopped the moment $\mathbf{Q}(\tau)$ comes to within the distance $\mathbf{r} + \delta_0$ from the border $\partial \mathcal{D}$ (here again $\mathbf{w}(\tau)$ is the standard two dimensional Brownian motion).

2.3. Heavy disk of small size.

We now turn to our last major result. The problems that plagued us in the previous theorem can be bypassed by taking the limit $\mathbf{r} \to 0$, in addition to $M \to \infty$. (That is we assume that the heavy particle is microscopically large but macroscopically small.) In this case the matrix $\sigma_Q^2(\mathcal{A})$ will be asymptotically constant, as we explain next. Recall that $\sigma_Q^2(\mathcal{A}) = \bar{\sigma}_Q^2(\mathcal{A})/\bar{L}$, where $\bar{L}$ does not depend on Q. Next, $\bar{\sigma}_Q^2(\mathcal{A})$ is given by the Green-Kubo formula (1.14). Its central term, corresponding to $j = 0$, can be found by a direct calculation:

$$(2.19) \qquad \int_{\Omega_Q} \mathcal{A}\mathcal{A}^T\, d\mu_Q = \frac{8\pi \mathbf{r}}{3\,(\mathrm{length}(\partial \mathcal{D}) + \mathrm{length}(\partial \mathcal{P}))}\, I,$$

see Section A.6, and it is independent of Q. Hence, the dependence of $\sigma_Q^2(\mathcal{A})$ on Q only comes from the correlation terms $j \neq 0$ in the series (1.14). Now, when the size of the massive disk is comparable to the size of the domain $\mathcal{D}$, the average time between its successive collisions with the light particle is of order one, and so those collisions are strongly correlated. By contrast, if $\mathbf{r} \approx 0$, the average time between successive interparticle collisions is $\mathcal{O}(1/\mathbf{r})$, and these collisions become almost independent. Thus, in the Green-Kubo formula (1.14), the central term (2.19) becomes dominant, and we arrive at

$$(2.20) \qquad \sigma_Q^2(\mathcal{A}) = \frac{8\mathbf{r}}{3\,\mathrm{Area}(\mathcal{D})}\, I + o(\mathbf{r}).$$

see Section A.6 for a complete proof.

Next, the time scale introduced in the previous theorem has to be adjusted to the present case where $\mathbf{r} \to 0$. Due to (1.18) and (2.20), we expect that the disk velocity grows as $\|V(t)\| = \mathcal{O}(\sqrt{\mathbf{r} t}/M)$, and its displacement as $\|Q(t) - Q_0\| = \mathcal{O}(t^{3/2} \mathbf{r}^{1/2}/M)$. Hence, for typical trajectories, it takes $\mathcal{O}(\mathbf{r}^{-1/3}M^{2/3})$ units of time before the heavy disk hits $\partial \mathcal{D}$. It is important that during this period of time $\|V(t)\| = \mathcal{O}(r^{1/3}M^{-2/3}) \approx 0$, hence $\|v\|$ remains close to one. We again modify the dynamics of the disk making it stop (freeze up) the moment it becomes δ_0-close to $\partial \mathcal{D}$ and consider the so modified dynamics of the heavy disk $Q_*(t), V_*(t)$ on time interval $(0, c\,\mathbf{r}^{-1/3}M^{2/3})$ with a constant $c > 0$. Our time scale is now

$$\tag{2.21} \tau = t\,\mathbf{r}^{1/3}M^{-2/3}$$

and we set

$$\tag{2.22} \mathcal{Q}(\tau) = Q_*(\tau\,\mathbf{r}^{-1/3}M^{2/3})$$

and hence

$$\tag{2.23} \mathcal{V}(\tau) = \mathbf{r}^{-1/3}M^{2/3}V_*(\tau\,\mathbf{r}^{-1/3}M^{2/3}).$$

It is easy to find an asymptotic distribution of $\mathcal{V}(\tau)$ by using our heuristic normal approximation (1.18) in a way similar to (2.5)–(2.7). Since the matrix $\sigma_Q^2(\mathcal{A})$ is almost constant, due to (2.20), we expect that $\mathcal{V}(\tau) \to \mathcal{N}(0, \sigma_0^2 \tau I)$, where

$$\tag{2.24} \sigma_0^2 = \frac{8}{3\,\mathrm{Area}(\mathcal{D})}.$$

The following theorem shows that our heuristic estimate is correct:

THEOREM 3. *Under the conditions A1–A3 there is a function $M_0 = M_0(\mathbf{r})$ such that if $\mathbf{r} \to 0$ and $M \to \infty$, so that $M > M_0(\mathbf{r})$, then the processes $\{\mathcal{Q}(\tau), \mathcal{V}(\tau)\}$ defined by (2.11), (2.21)–(2.23) converge weakly on the interval $0 < \tau < c$:*

$$\tag{2.25} \mathcal{V}(\tau) \to \sigma_0\,\mathbf{w}_{\mathcal{D}}(\tau)$$

and

$$\tag{2.26} \mathcal{Q}(\tau) \to Q_0 + \sigma_0 \int_0^\tau \mathbf{w}_{\mathcal{D}}(s)\,ds$$

where $\mathbf{w}_{\mathcal{D}}(\tau)$ is a standard two dimensional Brownian motion subjected to the following modification: we set $\mathbf{w}_{\mathcal{D}}(\tau) = 0$ for all $\tau > \tau_*$, where τ_* is the earliest moment when the right hand side of (2.26) becomes δ_0-close to $\partial \mathcal{D}$.

2.4. Comparison to previous works. There are two directions of research which our results are related to. The first one is the averaging theory of differential equations and the second is the study of long time behavior of mechanical systems.

The averaging theory deals with systems characterized by two types of dynamic variables, *fast* and *slow*. The case where the fast variables

2. STATEMENT OF RESULTS

make a Markov process, which does not depend on the slow variables, is quite well understood [**41**]. The results obtained in the Markov case have been extended to the situation where the fast motion is made by a hyperbolic dynamical system in [**55, 73, 35**]. By contrast, if the fast variables are coupled to the slow ones, as they are in our model, much less is known. Even the case where the fast motion is a diffusion process was settled quite recently [**71**]. Some results are available for coupled systems where the fast motion is uniformly hyperbolic [**2, 56**], but they all deal with relatively short time intervals, like the one in our relatively simple Theorem 1. The behavior during longer time periods, like those in our Theorems 2 and 3, remains virtually unexplored. This type of behavior is hard to control, it appears to be quite sensitive to the details of the problem at hand; for example, the uniqueness of the limiting process in our Theorem 2 relies upon the smoothness of the auxiliary function $\sigma_Q^2(\mathcal{A})$, cf. also Theorem 3 in [**71**].

Let us now turn to the second research field. While the phenomenological theory of the Brownian motion is more than a hundred years old (see [**39, 70**] for historic background) the mathematical understanding of how this theory can be derived from the microscopic Hamiltonian laws is still limited. Let us describe some available results refering the reader to the surveys [**22, 88, 93**] for more information.

Probably, the simplest mechanical model where one can observe a non-trivial statistical behavior is a periodic Lorentz gas. Bunimovich and Sinai [**9**] (see also an improved version in [**11**]) were the first to obtain a Brownian motion approximation for the *position* of a particle traveling in the periodic Lorentz gas with finite horizon. Their results hold for arbitrarily long intervals of time with respect to the equilibrium measure; such approximations are fairly common for chaotic dynamical systems [**33**]. On the contrary, we construct a Brownian motion approximation for relatively short periods of time, and in the context of Theorems 2–3 our system is far from equilibrium.

On the other hand, models of Brownian motion where one massive (tagged) particle is surrounded by an ideal gas of light particles have been studied in many papers, see, e.g., [**14, 36, 37, 48, 50, 87, 89, 92**]. We do not discuss these papers here referring the reader to the surveys [**88, 93**]. We observe that even though the methods of these papers do not play an important role in our proofs they should be useful for possible multiparticle extensions (see Section 9.5). The models in the above cited papers are more realistic in explaining the actual Brownian motion, however, there are still some unresolved questions. For example, it is commonly assumed that the gas is (and remains) in equilibrium, but there is no satisfactory mathematical explanation of why and how it reaches and maintains that state of equilibrium. (Moreover the results for out-of-equilibrium systems can differ from physical predictions made under the equilibrium assumption. See e.g. [94, 68].) We do not make (and do not need) such assumptions.

In fact, the equilibrium assumption is hard to substantiate. In ideal gases, where no direct interaction between gas particles takes place, equilibrium can only establish and propagate due to indirect interaction via collisions with the heavy particle and the walls. This process requires many collisions of each gas particle with the heavy one, but the existing techniques are incapable of tracing the dynamics beyond the time when each gas particle experiences just a few collisions, cf. [24, 25]. Our model contains a single light particle, but we are able to control the dynamics up to cM^a collisions (and actually longer, the main restriction of our analysis is the lack of control over $\sigma_Q(\mathcal{A})$ as the heavy disk approaches the border $\partial \mathcal{D}$, see Chapter 9). We hope that our method can be used to analyze systems of many particle as well.

We refer the reader to the surveys [47, 90, 88, 6] for descriptions of other models where macroscopic equations have been derived from deterministic microscopic laws. One of the main difficulties in deriving such equations is that microscopic equations of motion are time reversible, while the limit macroscopic equations are not, and therefore we cannot expect the convergence everywhere in phase space. Normally, the convergence occurs on a set of large (and, asymptotically, full) measure, which is said to represent "typical" phase trajectories of the system. In each problem, one has to carefully identify that large subset of the phase space and estimate its measure. If the system has some hyperbolic behavior, that large set has quite a complex fractal structure.

The existing approaches to the problem of convergence make use of certain families of measures on phase space, such that the convergence holds with probability ≈ 1 with respect to each of those measures. In the context of hyperbolic dynamical systems, the natural choice is the family of measures having smooth conditional distributions on unstable manifolds [85, 77, 74] (which is the characteristic property of Sinai-Ruelle-Bowen measures). Such measures work very well for averaging problems when the hyperbolicity is uniform and the dynamics is entirely smooth [55, 34, 35]. However, if the system has discontinuities and unbounded derivatives (as it happens in our case), the analysis of its behavior near the singularities becomes overwhelmingly difficult. Still, we will prove here that this general approach applies to systems with singularities.

CHAPTER 3

Plan of the proofs

3.1. General strategy. Our heuristic calculations of the asymptotic distribution of $\mathcal{V}(\tau)$ in Section 2.1 were based on the normal approximation (1.18), hence our main goal is to prove it. A natural approach is to fix the heavy disk at $Q = Q(t)$ and approximate the map $\mathcal{F}\colon \Omega \to \Omega$ by the billiard map $\mathcal{F}_{Q,V}\colon \Omega_{Q,V} \to \Omega_{Q,V}$, which is known [9, 96] to obey the central limit theorem (1.13). This approach, however, has obvious limitations.

On the one hand, our map $\mathcal{F}$ has positive Lyapunov exponents, hence its nearby trajectories diverge exponentially fast, so the above approximation (in a strict sense) only remains valid during time intervals $\mathcal{O}(\ln M)$, which are far shorter than we need. Thus, some sort of averaging is necessary to extend the CLT to longer time intervals. Here comes the second limitation: the central limit theorem for dispersing billiards (1.13) holds with respect to the billiard measure $\mu_{Q(t_0),V(t_0)}$, while we have to deal with the initial measure μ_{Q_0,V_0} and its images under our map $\mathcal{F}$, the latter might be quite different from $\mu_{Q(t_0),V(t_0)}$.

To overcome these limitations, we will show that the measures $\mathcal{F}^n(\mu_{Q_0,V_0})$ can be well approximated (in the weak topology) by averages (convex sums) of billiard measures $\mu_{Q,V}$:

$$(3.1) \qquad \mathcal{F}^n(\mu_{Q_0,V_0}) \sim \int \mu_{Q,V}\, d\lambda_n(Q,V)$$

where λ_n is some factor measure on the QV space. Furthermore, it is convenient to work with an even larger family of (auxiliary) measures, which we introduce shortly, and extend the approximation (3.1) to each auxiliary measure μ':

$$(3.2) \qquad \mathcal{F}^n(\mu') \sim \int \mu_{Q,V}\, d\lambda_n(Q,V)$$

for large enough n. In Section 3.3 below we make this approximation precise.

The proof of the 'equidistribution' (3.1)–(3.2) follows a shadowing type argument developed in the theory of uniformly hyperbolic systems without singularities [1, 34, 52, 78]. However, a major extra effort is required to extend this argument to systems with singularities, like ours. In fact, the largest error terms in our approximation (3.1) come from the orbits passing near singularities.

There are two places where we have trouble establishing (3.1)–(3.2) at all. First, if the velocity of the light particle becomes small, $\|v\| \approx 0$, then our system is no longer a small perturbation of a billiard dynamics (because the heavy disk can move a significant distance between successive collisions with the light particle). Second, if the heavy disk comes too close to the border $\partial \mathcal{D}$, then the mixing properties of the corresponding billiard dynamics deteriorate dramatically (roughly speaking, because the light particle can be trapped in a narrow tunnel between $\partial \mathcal{P}$ and $\partial \mathcal{D}$ for a long time). In this case the central limit theorem could only provide a satisfactory normal approximation to the billiard dynamics (in which the heavy disk is fixed) over very large times, but then the position and velocity of the heavy disk may change too much, rendering the billiard approximation itself useless.

Accordingly, we will fix a small $\delta_1 < \delta_0$, and most of the time we work in the region

$$(3.3) \quad (Q, V) \in \Upsilon_{\delta_1} := \{M\|V\|^2 < 1 - \delta_1, \quad \text{dist}(Q, \partial \mathcal{D}) > \mathbf{r} + \delta_1\}$$

(the first inequality guarantees that $\|v\| > \delta_1^{1/2} > 0$). We will show that violation of the first restriction (i.e. $\|v\| \leq \delta_1^{1/2}$) is improbable on the time scale we deal with, but possible violations of the other restriction (i.e. $\text{dist}(Q, \partial \mathcal{D}) \leq \mathbf{r} + \delta_1$) will force us to stop the heavy disk whenever it comes too close to the border $\partial \mathcal{D}$.

Our paper can be divided, roughly, into two parts, nearly equal in size but quite different in mathematical content. In the first, "dynamical" part (Chapters 4 and 5 and Appendices) we analyze the mechanical model of two particles, construct auxiliary measures, prove the equidistribution (3.1)–(3.2) and the log-Lipschitz continuity of the diffusion matrix (2.16). In the second, "probabilistic" part (Chapters 6–8) we prove the convergence to stochastic processes as claimed in Theorems 1–3; there we use various (standard and novel) moment-type techniques[1] [51, 31]. The arguments in Chapters 6–8 do not rely

---

[1] Note that the time scale in [31] corresponds to that of our Theorem 2. Indeed, in the notation of [31], $\varepsilon = 1/\sqrt{M}$ is the typical velocity of the heavy particle at equilibrium, hence their "4/3 law" becomes our "2/3 law". Let us also mention the papers [53, 54] studying the exit problem from a neighborhood of a "non-degenerate equilibrium" perturbed by a small noise. In these terms, our Theorem 2 deals with a "degenerate equilibrium", but the heuristic argument used to determine the correct scaling is similar to that of [53, 54].

on the specifics of the underlying dynamical systems, hence if one establishes results similar to (3.1)–(3.2) and (2.16) for another system, one would be able to derive analogues of our limit theorems by the same moment estimates.

The proofs of Theorems 1–3 follow similar lines, but Theorem 2 requires much more effort than the other two (mainly, because there are no explicit formulas for the limiting process, so we have to proceed in a roundabout way). We divide the proof of Theorem 2 between three sections: the convergence to equilibrium in the sense of (3.1)–(3.2) is established in Chapter 4, the log-Lipschitz continuity of the diffusion matrix (2.16) in Chapter 5, and the moment estimates specific to the scaling of Theorem 2 are done in Chapter 6. The modifications needed to prove the easier theorems 1 and 3 are described in Chapters 7 and 8, respectively.

3.2. Precise definitions. First we give the definition of auxiliary measures. Recall that our primary goal is control over measures $\mathcal{F}^n(\mu_{Q,V})$ for $n \geq 1$. The measure $\mu_{Q,V}$ is concentrated on the surface $\Omega_{Q,V}$. Let us coarse-grain this measure by partitioning $\Omega_{Q,V}$ into small subdomains $D \subset \Omega_{Q,V}$ and representing $\mu_{Q,V}$ as a sum of its restrictions to those domains. The image of a small domain $D \subset \Omega_{Q,V}$ under the map $\mathcal{F}^n$ gets strongly expanded in the unstable direction of the billiard map $\mathcal{F}_{Q,V}$, strongly contracted in the stable direction of $\mathcal{F}_{Q,V}$, slightly deformed in the transversal directions, and possibly cut by singularities of $\mathcal{F}$ into several pieces. Thus, $\mathcal{F}^n(D)$ looks like a union of one-dimensional curves that resemble unstable manifolds of the billiard map $\mathcal{F}_{Q,V}$, but may vary slightly in the transversal directions. Thus, the measure $\mathcal{F}^n(\mu_{Q,V})$ evolves as a weighted sum of smooth measures on such curves.

Motivated by this observation we introduce our family of auxiliary measures. A *standard pair* is $\ell = (\gamma, \rho)$, where $\gamma \subset \Omega$ is a C^2 curve, which is C^1 close to an unstable curve $\gamma_{Q,V} \subset \Omega_{Q,V}$ for the billiard map $\mathcal{F}_{Q,V}$ for some Q, V, and ρ is a smooth enough probability density on γ. The precise description of standard pairs is given in Chapter 4, here we only mention the properties of standard pairs most essential to our analysis. For a standard pair ℓ, we denote by γ_ℓ its curve, by ρ_ℓ its density, and by mes_ℓ the measure on γ with the density ρ_ℓ.

We define auxiliary measures via convex sums of measures on standard pairs, which satisfy an additional "length control":

Definition. An *auxiliary measure* is a probability measure m on Ω such that

$$(3.4) \qquad m = m_1 + m_2, \qquad |m_2| < M^{-50}$$

and m_1 is given by

$$(3.5) \qquad m_1 = \int \mathrm{mes}_{\ell_\alpha} \, d\lambda(\alpha)$$

where $\{\ell_\alpha = (\gamma_\alpha, \rho_\alpha)\}$ is a family of standard pairs such that $\{\gamma_\alpha\}$ make a measurable partition of Ω (m_1-mod 0), and λ is some factor measure satisfying

$$(3.6) \qquad \lambda\Big(\alpha \colon \, \mathrm{length}(\gamma_\alpha) < M^{-100}\Big) = 0$$

which imposes a "length control". We denote by $\mathfrak{M}$ the family of auxiliary measures.

It is clear that our family $\mathfrak{M}$ contains the initial smooth measure μ_{Q_0,V_0}, as well as every billiard measure $\mu_{Q,V}$ for Q,V satisfying $\|V\| < 1/\sqrt{M}$ and $\mathrm{dist}(Q, \partial \mathcal{D}) > \mathbf{r} + \delta_0/2$. Indeed, one can easily represent any of these measures by its conditional distributions on the fibers of a rather arbitrary smooth foliation of the corresponding space $\Omega_{Q,V}$ into curves whose tangent vectors lie in unstable cones (see precise definitions in Chapter 4).

Next we need to define a class of functions $\mathfrak{R} = \{A \colon \Omega \to \mathbb{R}\}$ satisfying two general (though somewhat conflicting) requirements. On the one hand, the functions $A \in \mathfrak{R}$ should be smooth enough on the bulk of the space Ω to ensure a fast (in our case – exponential) decay of correlations under the maps $\mathcal{F}_{Q,V}$. On the other hand, the regularity of the functions $A \in \mathfrak{R}$ should be compatible with that of the map $\mathcal{F}$, so that for any $A \in \mathfrak{R}$ the function $A \circ \mathcal{F}$ would also belong to $\mathfrak{R}$.

We will see in Chapter 4 that our map $\mathcal{F} \colon \Omega \to \Omega$ is not smooth, its singularity set $\mathcal{S}_1 = \partial \Omega \cup \mathcal{F}^{-1}(\partial \Omega)$ consists of points whose next collision is grazing. Due to our finite horizon assumption, $\mathcal{S}_1 \subset \Omega$ is a finite union of compact C^2 smooth submanifolds (with boundaries). We note that while the map $\mathcal{F}$ depends on the mass M of the heavy disk, its singularity set $\mathcal{S}_1$ does not. The complement $\Omega \setminus \mathcal{S}_1$ is a finite union of open connected domains, we call them Ω_k, $1 \leq k \leq k_0$. The restriction of the map $\mathcal{F}$ to each Ω_k is C^2. The derivatives of $\mathcal{F}$ are unbounded, but their growth satisfies the following inequality:

$$(3.7) \qquad \|D_x \mathcal{F}\| \leq L_\mathcal{F} \cdot [\mathrm{dist}(x, \mathcal{S}_1)]^{-1/2}$$

where $L_\mathcal{F} > 0$ is independent of M, see a proof in Section 4.1. In addition, the restriction of $\mathcal{F}$ to each Ω_k can be extended by continuity to the closure $\bar\Omega_k$, it then loses smoothness but remains Hölder continuous:

$$(3.8) \qquad \forall k \quad \forall x, y \in \bar\Omega_k \quad \|\mathcal{F}(x) - \mathcal{F}(y)\| \leq K_\mathcal{F} [\operatorname{dist}(x, y)]^{1/2}$$

where $K_\mathcal{F} > 0$ is independent of M, see a proof in Section 4.1.

These facts lead us to the following definition of $\mathfrak{R}$:

Definition. A function $A \colon \Omega \to \mathbb{R}$ belongs to $\mathfrak{R}$ iff
(a) A is continuous on $\Omega \setminus \mathcal{S}_1$. Moreover, the continuous extension of A to the closure of each connected component Ω_k of $\Omega \setminus \mathcal{S}_1$ is Hölder continuous with some exponent $\alpha_A \in (0, 1]$:

$$\forall k \quad \forall x, y \in \bar\Omega_k \quad |A(x) - A(y)| \leq K_A [\operatorname{dist}(x, y)]^{\alpha_A}$$

(b) at each point $x \in \Omega \setminus \mathcal{S}_1$ the function A has a local Lipschitz constant

$$(3.9) \qquad \operatorname{Lip}_x(A) := \limsup_{y \to x} |A(y) - A(x)| / \operatorname{dist}(x, y)$$

which satisfies the restriction

$$\operatorname{Lip}_x(A) \leq L_A [\operatorname{dist}(x, \mathcal{S}_1)]^{-\beta_A}$$

with some $L_A > 0$ and $\beta_A < 1$. The quantities $\alpha_A \leq 1$, $\beta_A < 1$, and $K_A, L_A > 0$ may depend on the function A.

Note that the set $\mathcal{S}_1$, and hence the class $\mathfrak{R}$, are independent of M. On the contrary, the singularity set

$$\mathcal{S}_n = \cup_{i=0}^n \mathcal{F}^{-1}(\partial\Omega)$$

of the map $\mathcal{F}^n$ depends on M for all $n \geq 2$. In Appendix B (Section B.3) we will prove the following:

LEMMA 3.1. *Let $n \geq 1$ and $B_1, B_2 \in \mathfrak{R}$. Then the function $A = B_1(B_2 \circ \mathcal{F}^{n-1})$ has the following properties:*
(a) *A is continuous on $\Omega \setminus \mathcal{S}_n$. Moreover, the continuous extension of A to the closure of each connected component $\Omega_{n,k}$ of the complement $\Omega \setminus \mathcal{S}_n$ is Hölder continuous with some exponent $\alpha_A \in (0, 1]$:*

$$\forall k \quad \forall x, y \in \bar\Omega_{n,k} \quad |A(x) - A(y)| \leq K_A [\operatorname{dist}(x, y)]^{\alpha_A}$$

(b) *at each point $x \in \Omega \setminus \mathcal{S}_n$ the local Lipschitz constant (3.9) of A satisfies the restriction*

$$\operatorname{Lip}_x(A) \leq L_A [\operatorname{dist}(x, \mathcal{S}_n)]^{-\beta_A}$$

with some $\beta_A < 1$. Here α_A, β_A, K_A, and L_A are determined by n and α_{B_i}, β_{B_i}, K_{B_i}, L_{B_i} for $i = 1, 2$, but they do not depend on M.

For any function $A\colon \Omega \to \mathbb{R}$ and a standard pair $\ell = (\gamma, \rho)$ we shall write
$$\mathbb{E}_\ell(A) = \int_\gamma A(x)\rho(x)\,dx.$$
We also define a projection $\pi_1(Q, V, q, v) = (Q, V)$ from Ω to the QV space.

3.3. Key technical results. With the above notation we are ready to state several propositions that give precise meaning to the heuristic formulas (3.1) and (3.2). According to (3.3), we will only deal with standard pairs $\ell = (\gamma, \rho)$ satisfying two restrictions:

(3.10) $\quad M\|\bar{V}^2\| \leq 1 - \delta_1 \quad \text{and} \quad \text{dist}(\bar{Q}, \partial\mathcal{D}) > \mathbf{r} + \delta_1$

for some $(\bar{Q}, \bar{V}) \in \pi_1(\gamma)$.

Next we fix a small $\delta_\diamond \ll \delta_1$ and will only deal with $\mathcal{F}^n$ where n satisfies $0 \leq n \leq \delta_\diamond\sqrt{M}$. This guarantees that the vital restrictions (3.10) will not be grossly violated, i.e. $\|v\|$ will stay close enough to a positive constant, cf. (1.4), and Q will stay away from the boundary $\partial\mathcal{D}$.

In all our propositions, K will denote sufficiently large constants, i.e. all our statements will hold true if $K > 0$ is large enough (the value of K can easily be chosen the same in all our estimates, so we will use the same plain symbol to avoid unnecessary indexation).

Our first proposition shows that the class of auxiliary measures is "almost" invariant under the dynamics (we can only claim "almost" invariance because (3.10) will eventually be violated).

PROPOSITION 3.2 (Propagation). *If $\ell = (\gamma, \rho)$ satisfies (3.10), then for all n satisfying*
$$K|\ln \text{length}(\gamma)| \leq n \leq \delta_\diamond\sqrt{M}$$
and any integrable function A we have

(3.11) $\quad \mathbb{E}_\ell(A \circ \mathcal{F}^n) = \sum_\alpha c_\alpha \mathbb{E}_{\ell_\alpha}(A)$

where $c_\alpha > 0$, $\sum_\alpha c_\alpha = 1$, and $\ell_\alpha = (\gamma_\alpha, \rho_\alpha)$ are standard pairs (the components of the image of ℓ under $\mathcal{F}^n$ with induced conditional measures); besides

(3.12) $\quad \displaystyle\sum_{\text{length}(\gamma_\alpha) < \varepsilon} c_\alpha \leq K\varepsilon$

for all $\varepsilon > 0$, the map $\mathcal{F}^{-n}$ is smooth on each γ_α, and

(3.13) $\quad \forall y', y'' \in \gamma_\alpha \quad \text{dist}[\mathcal{F}^{-m}(y'), \mathcal{F}^{-m}(y'')] \leq K\vartheta^m$

3. PLAN OF THE PROOFS

for all $1 \le m \le n$ and some constant $\vartheta \in (0,1)$.

The condition $K|\ln \text{length}(\gamma)| \le n$ is necessary to give short standard pairs enough time to expand and satisfy (3.12).

The next proposition basically shows that if ℓ satisfies (3.10), then

$$\mathcal{F}^n(\text{mes}_\ell) \approx \mu_{\bar{Q},\bar{V}}$$

(here we do not need to consider convex combinations yet) for all n in the range

(3.14) $\qquad \ln M \lesssim n \lesssim M^b, \qquad$ where $\quad 0 < b < 1/2$

The lower bound on n guarantees that $\mathcal{F}^n_{\bar{Q},\bar{V}}(\text{mes}_\ell)$ is almost uniformly distributed ('equidistributed') in $\Omega_{\bar{Q},\bar{V}}$, and the upper bound on n prevents Q and V from changing significantly during n iterations (implying that $\mathcal{F}^n_{\bar{Q},\bar{V}}$ will be still a good approximation to $\mathcal{F}^n$). We will use functions $A \colon \Omega \to \mathbb{R}$ such that

(3.15) $\qquad A = B_1\,(B_2 \circ \mathcal{F}^{n_A-1}), \qquad B_1, B_2 \in \mathfrak{R}$

for some small fixed $n_A \ge 1$ (independent of M). All the constants denoted by K will now depend on n_A as well.

PROPOSITION 3.3 (Short term equidistribution). *Let $\ell = (\gamma, \rho)$ satisfy (3.10) and A satisfy (3.15). Then for all n satisfying*

$$K|\ln \text{length}(\gamma)| \le n \le \delta_\diamond \sqrt{M}$$

and all $m \le \min\{n/2, K \ln M\}$ we have

$$\mathbb{E}_\ell(A \circ \mathcal{F}^n) = \mu_{\bar{Q},\bar{V}}(A) + \mathcal{O}(\mathcal{R}_{n,m} + \theta^m),$$

where $\theta \in (0,1)$ is a constant and

$$\mathcal{R}_{n,m} = \|\bar{V}\|(n+m^2) + (n^2+m^3)/M.$$

Even though Proposition 3.3 is formulated for a wider range of n than given by (3.14), it will only be useful when (3.14) holds, otherwise the error terms will be too big.

If we want to extend Proposition 3.3 to n beyond the upper bound in (3.14) and still keep the error terms small, we will have to deal with possible significant variation of the coordinates Q and V over the set $\mathcal{F}^n(\gamma_\ell)$. That can be done by using convex combinations of measures $\mu_{Q,V}$ (in the spirit of (3.1)), but it will be sufficient for us to restrict the analysis to a simpler case of functions A whose average $\mu_{Q,V}(A)$ does not depend on Q or V.

COROLLARY 3.4 (Long term equidistribution). *Let $\ell = (\gamma, \rho)$ satisfy (3.10), $\mathcal{A}$ satisfy (3.15), and, additionally, $\bar{A} = \mu_{Q,V}(\mathcal{A})$ be independent of Q, V. If*
$$K \,|\ln \operatorname{length}(\gamma)| \leq n \leq \delta_\diamond \sqrt{M},$$
then for all j satisfying
$$K \,|\ln \operatorname{length}(\gamma)| \leq j \leq n - K \,|\ln \operatorname{length}(\gamma)|$$
and $m \leq \min\{j/2, K \ln M\}$ we have
(3.16) $$\mathbb{E}_\ell(\mathcal{A} \circ \mathcal{F}^n) = \bar{A} + \mathcal{O}(\mathcal{R}_{n,j,m} + \theta^m),$$
where
$$\mathcal{R}_{n,j,m} = \mathbb{E}_\ell\left(\|V_{n-j}\|\right)(j + m^2) + (j^2 + m^3)/M,$$
and V_{n-j} denotes the V component of the point $\mathcal{F}^{n-j}(x)$, $x \in \gamma$.

Even though Corollary 3.4 is formulated for a wide range of j and m, it will only be useful when
$$\ln M \lesssim j, m \lesssim M^b, \qquad \text{where} \quad 0 < b < 1/2,$$
otherwise the error terms become too big. But the main number of iterations, n, can well grow up to $\delta_\diamond \sqrt{M}$, in this sense the corollary describes 'long term equidistribution'. To derive Corollary 3.4 we apply Proposition 3.2 with $n - j$ in place of n and then apply Proposition 3.3 (with j in place of n) to each α in (3.11), see Remark in the end of Chapter 4.

Lastly we state one more technical proposition necessary for the proof of Theorem 2. We formulate it in probabilistic terms (however, it is known to be equivalent to the fact that limiting factor measure $\lambda_n(Q, V)$ of (3.1) satisfies an associated partial differential equation):

PROPOSITION 3.5. (a) *Let $M_0 > 0$ and $a > 0$. The families of random processes $Q_*(\tau M^{2/3})$ and $M^{2/3} V_*(\tau M^{2/3})$ such that $M \geq M_0$, and the initial condition $(Q_0, V_0, q(0), v(0))$ is chosen randomly with respect to a measure in $\mathfrak{M}$ such that almost surely $\|V_0\| \leq a M^{-2/3}$, are tight and any limit process $(\mathbf{Q}(\tau), \mathbf{V}(\tau))$ satisfies (2.18).*
(b) *If the matrix $\sigma_Q(\mathcal{A})$ satisfies (2.16), then the equations (2.18) are well posed in the sense that any two solutions with the same initial conditions have the same distribution.*

The proofs of Propositions 3.2 and 3.3 and Corollary 3.4 are given in Chapter 4. The heart of the proof is contained in Sections 4.5 and 4.6, whereas Sections 4.1–4.4 extend some known results for classical billiards to our two-particle model. We remark that the estimates in Propositions 3.3 and 3.4 are likely to be less than optimal, but they suffice for our purposes because we restrict our analysis to time periods $\mathcal{O}(M^{2/3})$, which is much shorter than the ergodization time (the latter is apparently of order M, as one can see via a heuristic analysis similar to that in Section 1.3). Therefore to investigate the long time behavior of our system, the estimates of Propositions 3.3 and 3.4 might have to be sharpened (see Section 9.2), but we do not pursue this goal here.

Proposition 3.5 is proved in Chapter 6. In Chapters 7 and 8 we describe the modifications needed to prove Theorems 1 and 3 respectively. Chapter 7 is especially short since the material there is quite similar to [**35**, Sections 13 and 14], except that here some additional complications are due to the fact that we have to deal with a continuous time system.

CHAPTER 4

Standard pairs and equidistribution

The main goals of this section are the construction of standard pairs and the proofs of Statements 3.2, 3.3 and 3.4.

4.1. Unstable vectors. Our analysis will be restricted to the region (3.3). We first discuss the flow Φ^t in the full (seven-dimensional) phase space $\mathcal{M}$ in order to collect some preliminary estimates.

Let $x = (Q, V, q, v) \in \mathcal{M}$ be an arbitrary point and
$$dx = (dQ, dV, dq, dv) \in \mathcal{T}_x\mathcal{M}$$
a tangent vector. Let $dx(t) = D\Phi^t(dx)$ be the image of dx at time t. We describe the evolution of $dx(t)$ for $t > 0$.

Between successive collisions, the velocity components dV and dv remain unchanged, while the position components evolve linearly:

(4.1) $\quad dQ(t+s) = dQ(t) + s\, dV(t), \qquad dq(t+s) = dq(t) + s\, dv(t)$

At collisions, the tangent vector $dx(t)$ changes discontinuously, as we describe below.

First, we need to introduce convenient notation. For any unit vector $n \in \mathbb{R}^2$ (usually, a normal vector to some curve), we denote by $\mathbf{P}_n$ the projection onto n, i.e. $\mathbf{P}_n(u) = \langle u, n \rangle n$, and by $\mathbf{P}_n^\perp$ the projection onto the line perpendicular to n, i.e. $\mathbf{P}_n^\perp(u) = u - \mathbf{P}_n(u)$. Also, $\mathbf{R}_n$ denotes the reflection across the line perpendicular to n, that is
$$\mathbf{R}_n(u) = -\mathbf{P}_n(u) + \mathbf{P}_n^\perp(u) = u - 2\langle u, n \rangle n$$

For any vector $w \neq 0$, we write $\mathbf{P}_w^\perp$ for $\mathbf{P}_{w/\|w\|}^\perp$, for brevity.

Now, consider a collision of the light particle with the wall $\partial \mathcal{D}$, and let n denote the inward unit normal vector to $\partial \mathcal{D}$ at the point of collision. The components dQ and dV remain unchanged because the heavy disk is not involved in this event. The basic rule of specular reflection at $\partial \mathcal{D}$ reads $v^+ = \mathbf{R}_n(v^-)$ (the superscripts "+" and "−" refer to the postcollisional and precollisional vectors, respectively). Note that $\|v^+\| = \|v^-\|$. Accordingly, the tangent vectors dq and dv change by
$$dq^+ = \mathbf{R}_n(dq^-)$$

and
$$dv^+ = \mathbf{R}_n(dv^-) + \Theta^+(dq^+)$$
where
$$\Theta^+ = \frac{2\mathcal{K}\|v^+\|^2}{\langle v^+, n\rangle} \mathbf{P}^\perp_{v^+}$$

Here $\mathcal{K} > 0$ denotes the curvature of the boundary $\partial \mathcal{D}$ at the point of collision. Note that $\|dq^+\| = \|dq^-\|$. Also, $\Theta^+(dq^+) = \Theta^-(dq^-)$ where

$$\Theta^- = \frac{2\mathcal{K}\|v^+\|^2}{\langle v^+, n\rangle} \mathbf{R}_n \circ \mathbf{P}^\perp_{v^-} \tag{4.2}$$

Also, the geometry of reflection implies $\langle v^+, n\rangle > 0$.

Next, consider a collision between the two particles. At the moment of collision we have $q \in \partial \mathcal{P}(Q)$, i.e. $\|q - Q\| = \mathbf{r}$. Let $n = (q - Q)/\mathbf{r}$ be the normalized relative position vector. Then the laws of elastic collision (1.1)–(1.2) can be written as

$$\begin{aligned}
v^+ &= v^- - \frac{2M}{M+1}\mathbf{P}_n(v^- - V^-) \\
&= \mathbf{R}_n(v^-) + \frac{2M}{M+1}\left(\frac{1}{M}\mathbf{P}_n(v^-) + \mathbf{P}_n(V^-)\right) \\
V^+ &= V^- + \frac{2}{M+1}\mathbf{P}_n(v^- - V^-)
\end{aligned}$$

Let $w = v - V$ denote the relative velocity vector, cf. (1.5). Then
$$w^+ = w^- - 2\mathbf{P}_n(w^-) = \mathbf{R}_n(w^-)$$
and hence $\|w^+\| = \|w^-\|$. The components dq and dQ of the tangent vector dx change according to

$$dq^+ = \mathbf{R}_n(dq^-) + \frac{2M}{M+1}\left(\frac{1}{M}\mathbf{P}_n(dq^-) + \mathbf{P}_n(dQ^-)\right) \tag{4.3}$$

$$\begin{aligned}
dQ^+ &= \mathbf{R}_n(dQ^-) + \frac{2M}{M+1}\left(\frac{1}{M}\mathbf{P}_n(dq^-) + \mathbf{P}_n(dQ^-)\right) \\
&= dQ^- + \frac{2}{M+1}\mathbf{P}_n(dq^- - dQ^-)
\end{aligned} \tag{4.4}$$

Note that
$$dq^+ - dQ^+ = \mathbf{R}_n(dq^- - dQ^-)$$

and so $\|dq^+ - dQ^+\| = \|dq^- - dQ^-\|$. Next, the components dv and dV of the tangent vector dx change by

$$dv^+ = \mathbf{R}_n(dv^-) + \frac{2M}{M+1}\left(\frac{1}{M}\mathbf{P}_n(dv^-) + \mathbf{P}_n(dV^-)\right)$$
$$+ \frac{M}{M+1}\Theta^+(dq^+ - dQ^+)$$

and

$$dV^+ = dV^- + \frac{2}{M+1}\mathbf{P}_n(dv^- - dV^-)$$
$$- \frac{1}{M+1}\Theta^+(dq^+ - dQ^+)$$

where

$$\Theta^+ = \frac{2\mathcal{K}\|w^+\|^2}{\langle w^+, n\rangle}\mathbf{P}^\perp_{w^+}$$

Here $\mathcal{K} = 1/\mathbf{r}$ is the curvature of $\partial\mathcal{P}(Q)$. Note that $\Theta^+(dq^+ - dQ^+) = \Theta^-(dq^- - dQ^-)$, where

$$\Theta^- = \frac{2\mathcal{K}\|w^+\|^2}{\langle w^+, n\rangle}\mathbf{R}_n \circ \mathbf{P}^\perp_{w^-}$$

Also, the geometry of collision implies $\langle w^+, n\rangle > 0$, since we have chosen n to point *toward* the light particle.

All the above equations can be verified directly. Alternatively, one can use the fact that the system of two particles of different masses $M \neq m$ reduces to a billiard in a four dimensional domain by the change of variables $\tilde{Q} = Q\sqrt{M}$, $\tilde{V} = V\sqrt{M}$, $\tilde{q} = q\sqrt{m}$, and $\tilde{v} = v\sqrt{m}$ (the latter two are trivial since $m = 1$). This reduction is standard [84], and then the above equations can be derived from the general theory of billiards [16, 63, 84]. We omit the proof of the above estimates.

Now, since the total kinetic energy is fixed (1.3), the velocity components dv and dV of the tangent vector dx satisfy

(4.5) $$\langle v, dv\rangle + M\langle V, dV\rangle = 0$$

In addition, the Hamiltonian character of the dynamics implies that if the identity

(4.6) $$\langle v, dq\rangle + M\langle V, dQ\rangle = 0$$

holds at some time, it will be preserved at all times (future and past). From now on, we assume that all our tangent vectors satisfy (4.6).

There is a class of tangent vectors, which we will call *unstable vectors*, that is invariant under the dynamics. It is described in the following proposition:

PROPOSITION 4.1. *The class of tangent vectors dx with the following properties remains invariant under the forward dynamics:*

(a) $\langle dq, dv \rangle \geq (1 - C^{-1}) \|dq\| \|dv\|$
(b) $\|dQ\| \leq \frac{C}{M} \|dq\|$
(c) $\|dV\| \leq \frac{C}{M} \|dv\|$
(d) $\langle dq, v \rangle \leq C\|V\| \|dq\| \leq \frac{C}{\sqrt{M}} \|dq\|$
(e) $\langle dv, v \rangle \leq C\|V\| \|dv\| \leq \frac{C}{\sqrt{M}} \|dv\|$
(f) $\|dq\| \leq C \|dv\|$
(g) $\|dv\| \leq \frac{C}{|t^-(x)|} \|dq\|$
(h) *Equations (4.5) and (4.6) hold.*

Here $C > 1$ is a large constant, and $t^-(x) = \max\{t \leq 0 \colon \Phi^t(x) \in \Omega\}$ is the time of the latest collision along the past trajectory of x.

The proof of this proposition is based on the previous equations and some routine calculations, which we omit. □

We emphasize that our analysis has been done in the region (3.3) only, hence the above invariance holds as long as the system stays in Υ_{δ_1}; the constant C here depends on the choice of $\delta_1 > 0$, and we expect $C \to \infty$ as $\delta_1 \to 0$.

We also note that though unstable vectors make a multidimensional cone in the tangent space to Ω, this cone is essentially one-dimensional, its 'opening' in the Q and V directions is $\mathcal{O}(1/M)$. In the limit $M \to \infty$ we simply obtain the one-dimensional unstable cone for the classical billiard map.

Unstable vectors have strong (uniform in time) expansion property:

PROPOSITION 4.2. *Let dx be an unstable tangent vector and $dx(t) = D\Phi^t(dx)$ its image at time $t > 0$. Then the norm $\|dx(t)\|$ monotonically grows with t. Furthermore, there is a constant $\vartheta < 1$ such that for any two successive moments of collisions $t < t'$ of the light particle with $\partial \mathcal{D} \cup \partial \mathcal{P}(Q)$ we have*

(4.7) $$\|dx(t+0)\| \leq \vartheta \|dx(t'+0)\|$$

The notation $t+0$, $t'+0$ refer to the postcollisional vectors.

The proof easily follows from the previous equations. In fact,

(4.8) $$\vartheta^{-1} = 1 + L_{\min} \mathcal{K}_{\min}$$

where $\mathcal{K}_{\min} > 0$ is the smaller of $1/\mathbf{r}$ and the minimal curvature of $\partial \mathcal{D}$, $L_{\min}$ is the smaller of the minimal distance between the scatterers and δ_1, the minimal distance from the heavy disk to the scatterers allowed by (3.3).

At the moments of collisions it is more convenient (for technical reasons) to use the vector w defined by (1.5), instead of v, and respectively dw instead of dv. Then the vector w changes by the same rule $w^+ = \mathbf{R}_n(w^-)$ for both types of collisions (at $\partial \mathcal{D}$ and $\partial \mathcal{P}(Q)$). At collisions with $\partial \mathcal{D}$, the vector $dw = dv$ will change by the rule

$$dw^+ = \mathbf{R}_n(dw^-) + \Theta^+(dq^+)$$

while at collisions with the heavy disk, the vector $dw = dv - dV$ will change by a similar rule

$$dw^+ = \mathbf{R}_n(dw^-) + \Theta^+(dq^+ - dQ^+)$$

Furthermore, the expressions for Θ^+ and Θ^- will be identical for both types of collisions. The geometry of collision implies $\langle w, n \rangle \geq 0$ for both types of collisions.

It is easy to see that the inequalities (a)–(g) in Proposition 4.1 remain valid if we replace v by $v - V$ and dv by $dv - dV$ at any phase point, hence they apply to the vectors w and dw at the points of collision. The inequality (4.7) will also hold in the norm on Ω defined by

$$(4.9) \qquad \|dx\|^2 = \|dQ\|^2 + \|dV\|^2 + \|dq\|^2 + \|dw\|^2$$

Remark. Our equations show that the postcollisional tangent vector

$$(dQ^+, dV^+, dq^+, dw^+)$$

depends on the precollisional vector

$$(dQ^-, dV^-, dq^-, dw^-)$$

smoothly, unless $\langle w^+, n \rangle = 0$. This is the only singularity of the dynamics, it corresponds to grazing collisions (also, colloquially, called "tangential collisions").

Remark. Our equations imply that the derivative of the collision map $\mathcal{F}$ defined in Section 1.3 is bounded by $\|D_x \mathcal{F}\| \leq \text{Const}/\langle w^+, n \rangle$, where Const does not depend on M. It is easy to see that $\text{dist}(x, \mathcal{S}_1) = \mathcal{O}(\langle w^+, n \rangle^2)$, hence we obtain (3.7). Now (3.8) easily follows by integrating (3.7).

4.2. Unstable curves. We call a smooth curve $\mathcal{W} \subset \mathcal{M}$ an *unstable curve* (or a *u-curve*, for brevity) if, at every point $x \in \mathcal{W}$, the tangent vector to $\mathcal{W}$ is an unstable vector. By Propositions 4.1 and 4.2 the future image of a u-curve is a u-curve, which may be only piecewise smooth, due to singularities, and every u-curve is expanded by Φ^t monotonically and exponentially fast in time.

Now we extend our analysis to the collision map $\mathcal{F}\colon \Omega \to \Omega$. For every point $x \in \mathcal{M}$ we denote by $t^+(x) = \min\{t \geq 0\colon \Phi^t(x) \in \Omega\}$ and $t^-(x) = \max\{t \leq 0\colon \Phi^t(x) \in \Omega\}$ the first collision times in the future and the past, respectively. Let $\tilde{\pi}^\pm(x) = \Phi^{t^\pm(x)}(x) \in \Omega$ denote the respective "first collision" projection of $\mathcal{M}$ onto Ω. Note that $\mathcal{F}(x) = \tilde{\pi}^+(\Phi^\varepsilon x)$ for all $x \in \Omega$ and small $\varepsilon > 0$.

For any unstable curve $\mathcal{W} \subset \mathcal{M}$, the projection $W = \tilde{\pi}^-(\mathcal{W})$ is a smooth or piecewise smooth curve in Ω, whose components we also call *unstable curves* or *u-curves*. Let $dx = (dQ, dV, dq, dw)$ be the postcollisional tangent vector to $\mathcal{W}$ at a moment of collision (we remind the reader that $\dim \mathcal{M} = 7$ and $\dim \Omega = 6$). Its projection under the derivative $D\tilde{\pi}^-$ is a tangent vector $dx' = (dQ', dV', dq', dw')$ to the u-curve $W = \tilde{\pi}^-(\mathcal{W}) \subset \Omega$. Observe that $dV' = dV$, $dw' = dw$, $dQ' = dQ - tV$, and $dq' = dq - tv$, where t is uniquely determined by the condition $dx' \in \mathcal{T}_x\Omega$. Now some elementary geometry and an application of Proposition 4.1 give

PROPOSITION 4.3. *There is a constant* $1 < C < \infty$ *such that* $\|dQ'\| \leq C\|dv\|/\sqrt{M}$ *and* $C^{-1}\|dv\| \leq \|dq'\| \leq C\|dv\|$. *Therefore,*
$$\|dx'\|^2 = \left[\|dq'\|^2 + \|dv'\|^2\right]\left[1 + \mathcal{O}(1/\sqrt{M})\right],$$
and $C^{-1} \leq \|dx'\|/\|dx\| \leq C$.

We introduce two norms (metrics) on u-curves $W \subset \Omega$. First, we denote by length$(\cdot)$ the norm on W induced by the Euclidean norm $\|dx'\|$ on $\mathcal{T}_x(\Omega)$. Second, if $W = \tilde{\pi}^-(\mathcal{W})$, we denote by $|\cdot|$ the norm on W induced by the norm (4.9) on the postcollisional tangent vectors dx to $\mathcal{W}$ at the moment of collision. Due to the last proposition, these norms are equivalent in the sense

(4.10) $$C^{-1} \leq \frac{\text{length}(W)}{|W|} \leq C$$

By (4.10), we can replace length(γ) with $|\gamma|$ in the assumptions of Propositions 3.2 and 3.3, as well as in many other estimates of our paper. We actually prefer to work with the $|\cdot|$-metric, because it has an important *uniform expansion* property: the map $\mathcal{F}$ expands every u-curve in the $|\cdot|$-metric by a factor $\geq \vartheta^{-1} > 1$, see (4.8) (while the length$(\cdot)$ metric lacks this property).

Observe that the Q, V coordinates vary along u-curves $W \subset \Omega$ very slowly, so that u-curves are almost parallel to the cross-sections $\Omega_{Q,V}$ of Ω defined by (1.8). Each $\Omega_{Q,V}$ can be supplied with standard coordinates. Let r be the arc length parameter along $\partial \mathcal{D} \cup \partial \mathcal{P}(Q)$ and $\varphi \in [-\pi/2, \pi/2]$ the angle between the outgoing relative velocity vector

4. STANDARD PAIRS

$\bar{w}$ and the normal vector n. The orientation of r and φ is shown in Fig. 1. Topologically, $\Omega_{Q,V}$ is a union of cylinders, in which the cyclic coordinate r runs over the boundaries of the scatterers and the disk $\partial \mathcal{P}(Q)$, and $\varphi \in [-\pi/2, \pi/2]$. We need to fix reference points on each scatterer and on $\partial \mathcal{P}(Q)$ in order to define r, and then the coordinate chart r, φ in Ω_Q will actually be the same for all Q, V. We denote by Ω_0 that unique r, φ coordinate chart.

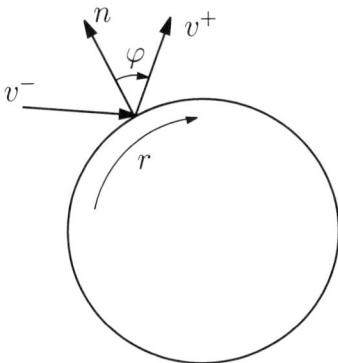

FIGURE 1. A collision of the light particle with a scatterer: the orientation of r and φ

Note that r and φ are defined at every point $x \in \Omega$, hence they make two coordinates in the (six-dimensional) space Ω. Since $\cos \varphi = \langle w, n \rangle / \|w\|$, the singularities of the map $\mathcal{F}$ correspond to $\cos \varphi = 0$, i.e. to $\varphi = \pm \pi/2$ (which is the boundary of Ω_0). Let π_0 denote the natural projection of Ω onto Ω_0. Note that, for each Q, V the projection $\pi_0 \colon \Omega_{Q,V} \to \Omega_0$ is one-to-one. Then the map

$$\pi_{Q,V} := \left(\pi_0|_{\Omega_{Q,V}}\right)^{-1} \circ \pi_0$$

defines a natural projection $\Omega \to \Omega_{Q,V}$ (geometrically, it amounts to moving the center of the heavy disk to Q, setting its velocity to V, and rescaling the vector w at points $q \in \partial \mathcal{P}(Q)$ by the rule (1.7)).

We turn back to u-curves $\mathcal{W} \subset \mathcal{M}$. For any such curve, $W = \pi_0(\tilde{\pi}^-(\mathcal{W}))$ is a smooth or piecewise smooth curve in Ω_0, whose components we also call *u-curves*. Any such curve is described by a smooth function $\varphi = \varphi(r)$. Let $dx = (dQ, dV, dq, dv)$ be the postcollisional tangent vector to $\mathcal{W}$ at a moment of collision. Its projection under the derivative $D(\pi_0 \circ \tilde{\pi}^-)$ is a tangent vector to W, which we denote by $(dr, d\varphi)$.

To evaluate $(dr, d\varphi)$, we introduce two useful quantities, $\mathcal{E}$ and $\mathcal{B}$, at each collision point. We set $\mathcal{E} = \|\mathbf{P}_w^{\perp}(dq)\|$ if the light particle collides with $\partial \mathcal{D}$, and $\mathcal{E} = \|\mathbf{P}_w^{\perp}(dq - dQ)\|$ if it collides with the disk. Then we set

$$\mathcal{B} = \frac{\|\mathbf{P}_w^{\perp}(dw)\|}{\mathcal{E}\,\|w\|}$$

PROPOSITION 4.4. *In the above notation,*

$$|dr| = \mathcal{E}/\cos\varphi \quad \text{and} \quad d\varphi/dr = \mathcal{B}\cos\varphi - \mathcal{K},$$

where $\mathcal{K} > 0$ is the curvature of $\partial \mathcal{D} \cup \partial \mathcal{P}$ at the point of collision. There is a constant $C > 1$ such that for any u-curve $W \subset \Omega$ and any point $x \in W$

$$\frac{2\mathcal{K}}{\cos\varphi} \leq \mathcal{B} \leq \frac{2\mathcal{K}}{\cos\varphi} + C$$

and

(4.11) $$C^{-1} \leq \frac{d\varphi}{dr} \leq C$$

In particular, $d\varphi/dr > 0$, hence $\pi_0(W)$ is an increasing curve in the r, φ coordinates. Lastly,

$$C^{-1} \leq \frac{(dr)^2 + (d\varphi)^2}{\|dx\|^2} \leq C$$

The proof is based on elementary geometric analysis, and we omit it. $\square$

Next we study the evolution of u-curves under the map $\mathcal{F}$. Let $W_0 \subset \Omega$ be a u-curve on which $\mathcal{F}^n$ is smooth for some $n \geq 1$. Then $W_i = \mathcal{F}^i(W_0)$ for $i \leq n$ are u-curves. Pick a point $x_0 \in W_0$ and put $x_i = \mathcal{F}^i(x_0)$ for $i \leq n$. For each i, we denote by r_i, φ_i, $\mathcal{K}_i$, $\mathcal{B}_i$, etc. the corresponding quantities, as introduced above, at the point x_i.

Also, for any u-curve $W \subset \Omega$ and $k \geq 1$ we denote by $\mathcal{J}_W \mathcal{F}^k(x)$ the Jacobian of the map $\mathcal{F}^k \colon W \to \mathcal{F}^k(W)$ at the point $x \in W$ in the norm $|\cdot|$, i.e. the local expansion factor of the curve W under $\mathcal{F}^k$ in the $|\cdot|$-metric.

PROPOSITION 4.5. *There is a constant $1 < C < \infty$ such that*

$$1 + \frac{C^{-1}}{\cos\varphi_{i+1}} < \mathcal{J}_{W_i}\mathcal{F}(x_i) < 1 + \frac{C}{\cos\varphi_{i+1}}$$

and

(4.12) $$\mathcal{J}_{W_0}\mathcal{F}^n(x_0) = \mathcal{J}_{W_0}\mathcal{F}(x_0) \cdots \mathcal{J}_{W_{n-1}}\mathcal{F}(x_{n-1}) \geq \vartheta^{-n}$$

where $\vartheta < 1$ is given by (4.8).

Proof. This follows from Proposition 4.4 by direct calculation. □

Remark. By setting $M = \infty$ in all our results we obtain their analogues for the billiard-type dynamics in $\mathcal{D} \setminus \mathcal{P}(Q)$, in which the disk $\mathcal{P}(Q)$ is fixed and the light particle moves at a constant speed $\|\bar{w}\| = \mathbf{s}_V$ given by (1.7). Most of them are known in the studies of billiards. In particular, we recover standard definitions of unstable vectors and unstable curves for the billiard-type map $\mathcal{F}_{Q,V} : \Omega_{Q,V} \to \Omega_{Q,V}$. Proposition 4.4 implies that the $\|dx\|$ norm on $\Omega_{Q,V}$ becomes

$$\|dx\|^2 = \|dq\|^2 + \|dw\|^2$$
$$(4.13) \qquad = (dr\cos\varphi)^2 + \mathbf{s}_V^2 (d\varphi + \mathcal{K}\, dr)^2$$

In this norm, the map $\mathcal{F}_{Q,V}$ expands every unstable curve by a factor $\geq \vartheta^{-1} > 1$.

As usual, reversing the time (changing $\mathcal{F}_{Q,V}$ to $\mathcal{F}_{Q,V}^{-1}$) gives the definition of stable vectors and stable curves (or *s-curves* for brevity) in the space $\Omega_{Q,V}$. Those are decreasing in the r, φ coordinates and satisfy the bound

$$(4.14) \qquad -C < d\varphi/dr < -C^{-1} < 0$$

The corresponding norm on stable vectors/curves is defined on *precollisional* tangent vectors and is expressed by

$$(4.15) \qquad \|dx\|_{\text{stable}}^2 = (dr\cos\varphi)^2 + \mathbf{s}_V^2 (d\varphi - \mathcal{K}\, dr)^2$$

which differs from (4.13) by the sign before $\mathcal{K}$. In the norm (4.15), the map $\mathcal{F}_{Q,V}$ contracts every s-curve by a factor $\leq \vartheta < 1$.

4.3. Homogeneous unstable curves. To control distortions of u-curves by the map $\mathcal{F}$, we need to carefully partition the neighborhood of the singularity set $\partial\Omega = \{\cos\varphi = 0\}$ into countably many surrounding sections (shells). This procedure has been introduced in [11] and goes as follows. Fix a large $k_0 \geq 1$ and for each $k \geq k_0$ define two "homogeneity strips" in Ω_0

$$\mathbb{H}_k = \{(r,\varphi) : \pi/2 - k^{-2} < \varphi < \pi/2 - (k+1)^{-2}\}$$

and

$$\mathbb{H}_{-k} = \{(r,\varphi) : -\pi/2 + (k+1)^{-2} < \varphi < -\pi/2 + k^{-2}\}$$

We also put

$$(4.16) \qquad \mathbb{H}_0 = \{(r,\varphi) : -\pi/2 + k_0^{-2} < \varphi < \pi/2 - k_0^{-2}\}$$

Slightly abusing notation, we will also denote by $\mathbb{H}_{\pm k}$ the preimages $\pi_0^{-1}(\mathbb{H}_{\pm k}) \subset \Omega$ and call them *homogeneity sections*. A u-curve $W \subset \Omega$

is said to be *weakly homogeneous* if W belongs to one section $\mathbb{H}_k$ for some $|k| \geq k_0$ or for $k = 0$.

Let $W \subset \mathbb{H}_k$ be a weakly homogeneous u-curve, $x = (r, \varphi) \in W$, and $|\Delta\varphi|$ be the projection of W onto the φ axis. Due to (4.11), we have

(4.17) $\quad |W| \leq \text{Const} \, |\Delta\varphi| \leq \text{Const} \, (|k|+1)^{-3} \leq \text{Const} \, \cos^{3/2} \varphi.$

Now let $W_0 \subset \Omega$ be a u-curve on which $\mathcal{F}^n$ is smooth, and assume that the u-curve $W_i = \mathcal{F}^i(W_0)$ is weakly homogeneous for every $i = 0, 1, \ldots, n$. Consider two points $x_0, x_0' \in W_0$ and put $x_i = \mathcal{F}^i(x_0)$ and $x_i' = \mathcal{F}^i(x_0')$ for $1 \leq i \leq n$. We denote by r_i, φ_i, $\mathcal{K}_i$, $\mathcal{B}_i$, etc. the corresponding quantities, as introduced in Section 4.2, at the point x_i, and by r_i', φ_i', $\mathcal{K}_i'$, $\mathcal{B}_i'$, etc. similar quantities at the point x_i'.

For any curve W we denote by $W(x, x')$ the segment of W between the points $x, x' \in W$ and by $\measuredangle(x, x')_W$ the angle between the tangent vectors to the curve W at x and x'.

PROPOSITION 4.6 (Distortion bounds). *Under the above assumptions, if the following bound holds for $i = 0$ with some $C_0 = c > 0$, then it holds for all $i = 1, \ldots, n-1$ with some $C_i = C > c$ (i.e., C_i is independent of i and n)*

$$\left| \ln \frac{\mathcal{J}_{W_i}\mathcal{F}(x_i)}{\mathcal{J}_{W_i}\mathcal{F}(x_i')} \right| \leq C_i \frac{|W_{i+1}(x_{i+1}, x_{i+1}')|}{|W_{i+1}|^{2/3}}$$

Moreover, in this case

$$\left| \ln \frac{\mathcal{J}_{W_0}\mathcal{F}^n(x_0)}{\mathcal{J}_{W_0}\mathcal{F}^n(x_0')} \right| \leq C \frac{|W_n(x_n, x_n')|}{|W_n|^{2/3}}$$

PROPOSITION 4.7 (Curvature bounds). *Under the above assumptions, if the following bound holds for $i = 0$ with some $C_0 = c > 0$, then it holds for all $i = 1, \ldots, n$ with some $C_i = C > c$ (independent of i and n)*

$$\measuredangle(x_i, x_i')_{W_i} \leq C_i \frac{|W_i(x_i, x_i')|}{|W_i|^{2/3}}$$

The proofs of these two propositions are quite lengthy. They are given in Appendix C. It is also shown there that sufficiently smooth unstable curves satisfy distortion bounds for a large enough $c > 0$. From now on we fix a sufficiently large $c > 0$ and the corresponding (perhaps even larger) $C > 0$ that guarantee the abundance of curves satisfying distortion and curvature bounds (this is a standard procedure in the study of chaotic billiards, see e.g. [**19**]).

We now consider an arbitrary u-curve $W \subset \Omega$ and partition it and its images under $\mathcal{F}^n$, $n \geq 1$, into weakly homogeneous u-curves (called H-components) as follows:

Definition (H-components). Given a u-curve $W \subset \Omega$, we call nonempty sets $W \cap \mathbb{H}_k$ (for $k = 0$ and $|k| \geq k_0$) the *H-components of W*. Note that W intersects each hyperplane $\{\varphi = \pm(\pi/2 - k^{-2})\}$ separating homogeneity sections at most once, due to (4.11), hence each H-component is a weakly homogeneous u-curve. Next suppose, inductively, that the H-components $W_{n,j}$, $j \geq 1$, of $\mathcal{F}^n(W)$ are constructed. Then the H-components of $\mathcal{F}^{n+1}(W)$ are defined to be the H-components of the u-curves $\mathcal{F}(W_{n,j})$ for all $j \geq 1$.

Observe that the H-components of $\mathcal{F}^n(W)$ are obtained naturally if we pretend that the boundaries of the homogeneity sections act as additional singularities of the dynamics.

Next, observe that if the curve W_0 satisfies the distortion bound and the curvature bound for $i = 0$, then so does any part of it (because $|W_0|$ and $|W_1|$ decrease if we reduce the size of the curve, thus the bounds in Propositions 4.6 and 4.7 remain valid). Therefore, if a weakly homogeneous u-curve W_0 satisfies the distortion bound and curvature bound for $i = 0$, then every H-component of its image $\mathcal{F}^n(W)$, $n \geq 1$ satisfies these bounds as well. This allows us to restrict our studies to weakly homogeneous u-curves that satisfy the distortion and curvature bounds:

Definition (H-curves). A weakly homogeneous u-curve W_0 is said to be *homogeneous* (or an *H-curve*, for brevity) if it satisfies the above distortion bound and curvature bound.

We note that, in the notation of Proposition 4.6,
$$|W_n(x_n, x'_n)|/|W_n|^{2/3} \leq |W_n|^{1/3} \leq \text{Const},$$
hence the distortions of H-curves under the maps $\mathcal{F}^n$, $n \geq 1$, are uniformly bounded, in particular, for some constant $\tilde{\beta} > 0$

(4.18) $\quad e^{-\tilde{\beta}} \dfrac{|W_0(x_0, x'_0)|}{|\mathcal{F}^{-n}(W_n)|} \leq \dfrac{|W_n(x_n, x'_n)|}{|W_n|} \leq e^{\tilde{\beta}} \dfrac{|W_0(x_0, x'_0)|}{|\mathcal{F}^{-n}(W_n)|}.$

Moreover, Proposition 4.7 implies

(4.19) $\quad\quad\quad \measuredangle(x, x')_W \leq e^{\tilde{\beta}} \quad\quad \forall x, x' \in W.$

Now consider an H-curve W_0 such that $W_i = \mathcal{F}^i(W_0)$ is an H-curve for every $i = 1, \ldots, n$. Let mes_0 be an absolutely continuous measure on W_0 with some density ρ_0 with respect to the measure induced by the $|\cdot|$-norm. Then $\text{mes}_i = \mathcal{F}^i(\text{mes}_0)$ is a measure on the curve W_i with some density ρ_i for each $i = 1, \ldots, n$. As an immediate consequence of Proposition 4.6 and (4.18), we have

COROLLARY 4.8 (Density bounds). *Under the above assumptions, if the following bound holds for $i = 0$ with some $C_0 = c > 0$, it holds for all $i = 1, \ldots, n$ with some $C_i = C > c$ (independent of i and n)*

$$\left| \ln \frac{\rho_i(x_i)}{\rho_i(x_i')} \right| \leq C_i \frac{|W_i(x_i, x_i')|}{|W_i|^{2/3}} \tag{4.20}$$

Observe that if the density bound holds for $i = 0$ on the curve W_0, then it holds on any part of it (because $|W_0|$ decreases if we reduce the size of the curves, so the bound (4.20) remains valid). Therefore, if W_0 is an arbitrary H-curve with a density ρ_0, then the map $\mathcal{F}^n$, $n \geq 1$, induces densities on the H-components of $\mathcal{F}^n(W_0)$ that satisfy the above density bound. Hence we can restrict our studies to densities satisfying (4.20):

Definition. Given an H-curve W_0, we say that ρ_0 is a *homogeneous density* if it satisfies (4.20).

Note that (4.20) remains valid whether we normalize the corresponding densities or not. Also, because $|W(x, x')|/|W|^{2/3} < |W|^{1/3} <$ Const, we have a uniform bound

$$e^{-\tilde{\beta}} \leq \frac{\rho(x)}{\rho(x')} \leq e^{\tilde{\beta}} \qquad \forall x, x' \in W, \tag{4.21}$$

where $\tilde{\beta} = C_0 \max_W |W|^{1/3}$.

We will require H-curves to have length shorter than a small constant $\tilde{\delta} > 0$ (to achieve this, large H-curves can be always partitioned into H-curves of length between $\tilde{\delta}/2$ and $\tilde{\delta}$), so that $\tilde{\beta}$ in (4.18), (4.21) and (4.19) is small enough. This will make our H-curves almost straight lines, the map $\mathcal{F}$ on them will be almost linear, and homogeneous densities will be almost constant.

4.4. Standard pairs. Now we formally define standard pairs mentioned earlier in Section 3.3:

Definition (Standard pairs). A standard pair $\ell = (\gamma, \rho)$ is an H-curve $\gamma \subset \Omega$ with a homogeneous probability density ρ on it. We denote by mes = mes_ℓ the measure on γ with density ρ.

Our previous results imply the following invariance of the class of standard pairs:

PROPOSITION 4.9. *Let $\ell = (\gamma, \rho)$ be a standard pair. Then for each $n \geq 0$, we have $\mathcal{F}^n(\gamma) = \cup_i \gamma_{i,n}$ and $\mathcal{F}^n(\mathrm{mes}_\ell) = \sum_i c_i \mathrm{mes}_{\ell_{i,n}}$ where $\sum_i c_i = 1$ and $\ell_{i,n} = (\gamma_{i,n}, \rho_{i,n})$ are standard pairs. The curves $\gamma_{i,n}$ are the H-components of $\mathcal{F}^n(\gamma)$. Furthermore, any subcurve $\gamma' \subset \gamma_{i,n}$ with the density $\rho_{i,n}$ restricted to it, is a standard pair.*

Recall that $\mathcal{F}$ expands H-curves by a factor $\geq \vartheta^{-1} > 1$, which is a local property. It is also important to show that H-curves grow in a global sense, i.e. given a small H-curve γ, the sizes of the H-components of $\mathcal{F}^n(\gamma)$ tend to grow exponentially in time until they become of order one, on the average (we will make this statement precise). Such statements are usually referred to as "growth lemmas" [11, 96, 18, 19], and we prove one below.

Let $\ell = (\gamma, \rho)$ be a standard pair and for $n \geq 1$ and $x \in \gamma$, let $r_n(x)$ denote the distance from the point $\mathcal{F}^n(x)$ to the nearest endpoint of the H-component $\gamma_n(x) \subset \mathcal{F}^n(\gamma)$ that contains $\mathcal{F}^n(x)$.

LEMMA 4.10 ("Growth lemma"). *If k_0 in (4.16) is sufficiently large, then*
(a) *There are constants $\beta_1 \in (0,1)$ and $\beta_2 > 0$, such that for any $\varepsilon > 0$*
(4.22) $\quad \mathrm{mes}_\ell(x \colon r_n(x) < \varepsilon) \leq (\beta_1/\vartheta)^n \mathrm{mes}_\ell(x \colon r_0 < \varepsilon \vartheta^n) + \beta_2 \varepsilon$
(b) *There are constants $\beta_3, \beta_4 > 0$, such that if $n \geq \beta_3 |\ln |\gamma||$, then for any $\varepsilon > 0$ we have $\mathrm{mes}_\ell(x \colon r_n(x) < \varepsilon) \leq \beta_4 \varepsilon$.*
(c) *There are constants $\beta_5, \beta_6 > 0$, a small $\varepsilon_0 > 0$, and $q \in (0,1)$ such that for any $n_2 > n_1 > \beta_5 |\ln |\gamma||$ we have*

$$\mathrm{mes}_\ell\left(x \colon \max_{n_1 < i < n_2} r_i(x) < \varepsilon_0\right) \leq \beta_6 q^{n_2 - n_1}$$

All these estimates are uniform in $\ell = (\gamma, \rho)$.

PROOF. The proof of (a) follows the lines of the arguments in [18] and consists of three steps.

First, let γ be an H-curve and $\gamma_{i,1}$ all the H-components of $\mathcal{F}(\gamma)$. For each i, denote by $\vartheta_{i,1}^{-1}$ the minimal (local) factor of expansion of the curve $\mathcal{F}^{-1}(\gamma_{i,1})$ under the map $\mathcal{F}$. We claim that

(4.23) $$\boldsymbol{\theta}_1 := \lim_{\delta \to 0} \sup_{\gamma \colon |\gamma| < \delta} \sum_i \vartheta_{i,1} < 1$$

(we call this a *one-step expansion estimate* for the map $\mathcal{F}$).

To prove (4.23), we observe that a small H-curve γ may be cut into several pieces by the singularities of $\mathcal{F}$, which are made by grazing (tangential) collisions with the scatterers and the disk $\mathcal{P}(Q)$. At each of them, γ is sliced into two parts – one hits the scatterer (or the disk) and gets reflected (almost tangentially) and the other misses the collision (passes by). The reflecting part is further subdivided into countably many H-components by the boundaries of the homogeneity sections $\mathbb{H}_k$. Note that the reflecting part of γ lies entirely in $\cup_{k \geq k_0} \mathbb{H}_{\pm k}$ provided $|\gamma|$ is small enough, which is guaranteed by taking $\liminf_{\delta \to 0}$.

Let γ' be an H-component of $\mathcal{F}(\gamma)$ falling into a section $\mathbb{H}_k$ with some $|k| \geq k_0$. Since $\cos \varphi \sim k^{-2}$ on γ', the expansion factor of the preimage $\mathcal{F}^{-1}(\gamma')$ under the map $\mathcal{F}$ is $\geq ck^2$ for some constant $c > 0$, due to Proposition 4.5. Thus all these H-components make a total contribution to (4.23) less than $\sum_{k \geq k_0} (ck^2)^{-1} \leq \text{Const}/k_0$.

The part of γ passing by without collision may be sliced again by a grazing collision with another scatterer later on, thus creating another countable set of reflecting H-components. This can happen at most $L_{\max}/L_{\min}$ times, where $L_{\min}$ is the minimal free path of the light particle, guaranteed by (3.3), and $L_{\max}$ is the maximal free path of the light particle ($L_{\max} < \infty$ due to our finite horizon assumption).

In the end, we will have $\leq L_{\max}/L_{\min}$ countable sets of H-components resulting from almost grazing collisions and at most one component of γ that misses all the grazing collisions and lands somewhere else on $\partial \mathcal{D} \cup \partial \mathcal{P}(Q)$. That last component is only guaranteed to expand by a moderate factor of ϑ^{-1}. Thus, we arrive at

$$(4.24) \qquad \boldsymbol{\theta}_1 \leq \vartheta + \frac{L_{\max}}{L_{\min}} \frac{\text{Const}}{k_0}$$

Since $\vartheta < 1$, the required condition $\boldsymbol{\theta}_1 < 1$ can be ensured by choosing k_0 large enough. This completes the proof of the one-step expansion estimate (4.23).

The second step in the proof of Lemma 4.10 (a) is the verification of (4.22) for $n = 1$:

$$(4.25) \qquad \text{mes}_\ell(x: \, r_1(x) < \varepsilon) \leq (\beta_1/\vartheta) \, \text{mes}_\ell(x: \, r_0 < \varepsilon \vartheta) + \beta_2 \varepsilon.$$

We assume that $|\gamma| < \tilde{\delta}$, where $\tilde{\delta}$ is chosen so that

$$\tilde{\boldsymbol{\theta}}_1 := \sup_{\gamma: \, |\gamma| < \tilde{\delta}} \sum_i \vartheta_{i,1} < (1 + \boldsymbol{\theta}_1)/2 < 1.$$

Now, for each H-component $\gamma_{i,1}$ of $\mathcal{F}(\gamma)$, the set $\gamma_{i,1} \cap \mathcal{F}(\{r_1(x) < \varepsilon\})$ is the union of two subintervals of $\gamma_{i,1}$ of length ε adjacent to the endpoints of $\gamma_{i,1}$. Then the set $\mathcal{F}^{-1}(\gamma_{i,1}) \cap \{r_1(x) < \varepsilon\}$ is a subset of the union of two subintervals of $\mathcal{F}^{-1}(\gamma_{i,1})$ of length $\vartheta_{i,1}\varepsilon$, therefore,

$$\text{mes}_\ell(r_1(x) < \varepsilon) \leq |\gamma|^{-1} e^{\tilde{\beta}} \sum_i 2\varepsilon \vartheta_{i,1}$$

$$\text{(4.26)} \qquad \leq 2\varepsilon |\gamma|^{-1} e^{\tilde{\beta}} \tilde{\boldsymbol{\theta}}_1$$

where the factor $e^{\tilde{\beta}}$ accounts for possible fluctuations of the density $\rho(x)$ on γ, see (4.21). We can make $\tilde{\beta} > 0$ arbitrarily small by decreasing $\tilde{\delta}$, if necessary, and guarantee that

$$\beta_1 := e^{2\tilde{\beta}} \tilde{\boldsymbol{\theta}}_1 < 1 \qquad \text{and} \qquad \beta_1/\vartheta > 1$$

(recall that the first bound is required by Lemma 4.10; the second one can be easily ensured because $\vartheta < \tilde{\boldsymbol{\theta}}_1 < 1$). Now the first term on the right hand side of (4.25) is bounded below by

$$(\beta_1/\vartheta) \text{mes}_\ell(x: r_0 < \varepsilon\vartheta) \geq (\beta_1/\vartheta) \min\{1, 2\varepsilon\vartheta |\gamma|^{-1} e^{-\tilde{\beta}}\}$$
$$= \min\{\beta_1/\vartheta, 2\varepsilon |\gamma|^{-1} e^{\tilde{\beta}} \tilde{\boldsymbol{\theta}}_1\}$$

Since $\beta_1/\vartheta > 1$, we obtain

$$\text{(4.27)} \qquad \text{mes}_\ell(x: r_1(x) < \varepsilon) \leq (\beta_1/\vartheta) \text{mes}_\ell(x: r_0 < \varepsilon\vartheta).$$

This bound appears even better than (4.25), but remember it is only proved under the assumption $|\gamma| < \tilde{\delta}$. To make this assumption valid, we require all our H-curves to have length shorter than $\tilde{\delta}$, as already mentioned in the end of the previous section. Accordingly, we have to partition the H-components of $\mathcal{F}(\gamma)$ into pieces that are shorter than $\tilde{\delta}$; this will enlarge the set $\{r_1(x) < \varepsilon\}$ and result in the additional term $\beta_2 \varepsilon$ in (4.25).

More precisely, let us divide each H-component $\gamma_{i,1}$ of $\mathcal{F}(\gamma)$ with length $> \tilde{\delta}$ into k_i equal subintervals of length between $\tilde{\delta}/2$ and $\tilde{\delta}$, with $k_i \leq 2|\gamma_{i,1}|/\tilde{\delta}$. If $|\gamma_{i,1}| \leq \tilde{\delta}$, then we set $k_i = 0$ and leave $\gamma_{i,1}$ unchanged. Then the union of the preimages of the ε-neighborhoods of the new partition points has measure bounded above by

$$\leq 3\varepsilon |\gamma|^{-1} \sum_i k_i \vartheta_{i,1} \leq 6\varepsilon \tilde{\delta}^{-1} |\gamma|^{-1} \sum_i |\gamma_{i,1}| \vartheta_{i,1} \leq 7\varepsilon \tilde{\delta}^{-1},$$

where we increased the numerical coefficient from 6 to 7 in order to incorporate the factor $e^{\tilde{\beta}}$ resulting from the distortion bounds (4.18). This completes the proof of (4.25) with $\beta_2 = 7\tilde{\delta}^{-1}$.

Lastly, the proof of (4.22) for $n > 1$ goes by induction on n. Assume that
$$\operatorname{mes}_\ell\big(x\colon r_n(x) < \varepsilon\big) \leq (\beta_1/\vartheta)^n \operatorname{mes}_\ell\big(x\colon r_0(x) < \varepsilon\vartheta^n\big)$$
$$+ 7\tilde{\delta}^{-1}\big(1 + \beta_1 + \cdots + \beta_1^{n-1}\big)\varepsilon.$$

Then we apply (4.25) with $\beta_2 = 7\tilde{\delta}^{-1}$ to each H-component of $\mathcal{F}^n(\gamma)$ and obtain
$$\operatorname{mes}_\ell\big(x\colon r_{n+1}(x) < \varepsilon\big) \leq (\beta_1/\vartheta) \operatorname{mes}_\ell\big(x\colon r_n(x) < \varepsilon\vartheta\big) + 7\tilde{\delta}^{-1}\varepsilon$$
$$\leq (\beta_1/\vartheta)^{n+1} \operatorname{mes}_\ell\big(x\colon r_0(x) < \varepsilon\vartheta^{n+1}\big)$$
$$+ 7\tilde{\delta}^{-1}\big(1 + \beta_1 + \cdots + \beta_1^n\big)\varepsilon,$$

which completes the induction step. Thus we get (4.22) for all $n \geq 1$ with $\beta_2 = 7\tilde{\delta}^{-1}/(1 - \beta_1)$.

Part (b) of Lemma 4.10 directly follows from (a). Indeed, it suffices to set $\beta_3 = 1/\min\{|\ln\beta_1|, |\ln\vartheta|\}$, so that $\vartheta^n < |\gamma|$ and $\beta_1^n < |\gamma|$, and notice that $\operatorname{mes}_\ell(x\colon r_0 < \varepsilon\vartheta^n) < 2e^{\tilde{\beta}}\varepsilon\vartheta^n/|\gamma|$ due to (4.21).

The proof of (c) requires a tedious bookkeeping of various short H-components of the images of γ. Pick $\varepsilon_0 < (1 + \beta_4)^{-1}$ and divide the time interval $[n_1, n_2]$ into segments of length $s\colon = [2\beta_3|\ln\varepsilon_0|]$. We will estimate the measure of the set
$$\tilde{\gamma} = \Big\{x \in \gamma\colon \max_{1 \leq i \leq K} r_{n_1+si}(x) < \varepsilon_0\Big\}$$
where $K = (n_2 - n_1)/s$. For each $x \in \tilde{\gamma}$ define a sequence of natural numbers $S(x) = \{k_0, k_1, \ldots, k_m\}$, with $m = m(x) \leq K$, inductively. Set $k_0 = 1$ and given $k_0, \ldots, k_i$ we put $t_i = k_0 + \cdots + k_i$ and consider the H-component $\gamma_i(x)$ of $\mathcal{F}^{n_1+st_i}(\gamma)$ that contains $\mathcal{F}^{n_1+st_i}(x)$. We set $k_{i+1} = k$ if $|\gamma_i(x)| \in [\varepsilon_0^{2k}, \varepsilon_0^{2k-2})$. If it happens that $t_i + k_{i+1} > K$, we reset $k_{i+1} = K - t_i + 1$ and put $m(x) = i + 1$. Note that now $k_1 + \cdots + k_m = K$.

Next pick a sequence $S = \{k_0 = 1, k_1, \ldots, k_m\}$ of natural numbers such that $k_1 + \cdots + k_m = K$ and let $\tilde{\gamma}_S = \{x \in \tilde{\gamma}\colon S(x) = S\}$. We claim that

(4.28) $$\operatorname{mes}_\ell(\tilde{\gamma}_S) \leq \beta_4^m \varepsilon_0^K$$

First, by part (b)

(4.29) $$\operatorname{mes}_\ell\big(k_1(x) = k\big) \leq \beta_4 \varepsilon_0^k, \qquad k \geq 1$$

(for $k \geq 2$ we actually have a better estimate $\operatorname{mes}_\ell(k_1(x) = k) \leq \beta_4 \varepsilon_0^{2k-2}$). Then, inductively, for each $i = 0, \ldots, m-2$ we use our previous notation t_i and $\gamma_i(x)$ and put $\tilde{\gamma}_i(x) = \gamma_i(x)$ if $|\gamma_i(x)| < 2\varepsilon_0$, otherwise we denote by $\tilde{\gamma}_i(x) \subset \gamma_i(x)$ the ε_0-neighborhood of an endpoint of $\gamma_i(x)$ that contains the point $\mathcal{F}^{n_1+st_i}(x)$. By Proposition 4.9, the curve $\tilde{\gamma}_i(x)$, with the corresponding conditional measure on it (induced by $\mathcal{F}^{n_1+st_i}(\text{mes}_\ell)$), makes a standard pair, call it $\ell_i(x)$. Then again by part (b)

$$(4.30) \qquad \text{mes}_{\ell_i(x)}\bigl(k_{i+2}(x) = k\bigr) \leq \beta_4 \varepsilon_0^k, \qquad k \geq 1$$

(because $k_{i+2}(x) = k$ implies $|\gamma_{i+1}(x)| < \varepsilon_0^{2k-2}$, which is enough for $k \geq 2$, and for $k = 1$ we have $|\tilde{\gamma}_{i+1}(x)| < \varepsilon_0$). Multiplying (4.29) and (4.30) for all $i = 0, \ldots, m-2$ proves (4.28).

Now, adding (4.28) over all possible sequences $S = \{1, k_1, \ldots, k_m\}$ gives

$$\text{mes}_\ell(\tilde{\gamma}) \leq \sum_{m=1}^{K} \binom{K-1}{m-1} \beta_4^m \varepsilon_0^K \leq (1+\beta_4)^K \varepsilon_0^K$$

where $\binom{K-1}{m-1}$ denotes the binomial coefficients coming from counting the number of respective sequences $\{S\}$. This completes the proof of Lemma 4.10. $\square$

Proof of Proposition 3.2 is now obtained by combining Proposition 4.9 with Lemma 4.10. $\square$

We conclude this subsection with a few remarks. Let $\gamma = \cup_\alpha \gamma_\alpha \subset \Omega$ be a finite or countable union of disjoint H-curves with some smooth probability measure mes_γ on it, whose density of each γ_α is homogeneous. For every α and $x \in \gamma_\alpha$ and $n \geq 0$ denote by $r_n(x)$ the distance from the point $\mathcal{F}^n(x)$ to the nearest endpoint of the H-component of $\mathcal{F}^n(\gamma_\alpha)$ to which the point $\mathcal{F}^n(x)$ belongs. The following is an easy consequence of Lemma 4.10 (a) obtained by averaging over α:

$$\text{mes}_\gamma\bigl(x\colon r_n(x) < \varepsilon\bigr) \leq (\beta_1/\vartheta)^n \text{mes}_\gamma\bigl(x\colon r_0 < \varepsilon \vartheta^n\bigr) + \beta_2 \varepsilon.$$

Also, there is a constant $\beta_7 > 0$ such that if $\text{mes}_\gamma(x\colon r_0(x) < \varepsilon) \leq \beta_7 \varepsilon$ for any $\varepsilon > 0$, then $\text{mes}_\gamma(x\colon r_n(x) < \varepsilon) \leq \beta_7 \varepsilon$ for all $\varepsilon > 0$ and $n \geq 1$ (it is enough to set $\beta_7 = \beta_2/(1 - \beta_1 e^{\tilde{\beta}})$).

In addition, suppose for each α we fix a subcurve $\gamma'_\alpha \subset \gamma_\alpha$. Put $\gamma' = \cup_\alpha \gamma'_\alpha$ and denote by $\text{mes}_{\gamma'}$ the measure mes_γ conditioned on γ'. For every α and $x \in \gamma'_\alpha$ and $n \geq 0$ denote by $r'_n(x)$ the distance from the point $\mathcal{F}^n(x)$ to the nearest endpoint of the H-component of $\mathcal{F}^n(\gamma'_\alpha)$ to which the point $\mathcal{F}^n(x)$ belongs. Distortion bounds (4.18) then imply

$$\text{mes}_{\gamma'}\bigl(x\colon r'_n(x) < \varepsilon\bigr) \leq e^{\tilde{\beta}} \bigl[\text{mes}_\gamma(\gamma')\bigr]^{-1} \text{mes}_\gamma\bigl(x\colon r_n(x) < \varepsilon\bigr)$$

Lastly, we note that taking the limit $M \to \infty$ automatically extends all our results to the billiard map $\mathcal{F}_Q \colon \Omega_Q \to \Omega_Q$ for any Q satisfying (3.3).

4.5. Perturbative analysis. Recall that the billiard-type map $\mathcal{F}_{Q,V} \colon \Omega_{Q,V} \to \Omega_{Q,V}$ is essentially independent of V and can be identified with $\mathcal{F}_Q \colon \Omega_Q \to \Omega_Q$ via $\pi_{Q,0} \circ \mathcal{F}_{Q,V} = \mathcal{F}_Q \circ \pi_{Q,0}$. Furthermore, the spaces Ω_Q are identified with the r, φ coordinate space Ω_0 by the projection π_0. This gives us a family of billiard maps $\mathcal{F}_Q$ acting on the same space Ω_0. They preserve the same billiard measure $d\mu_0 = c^{-1} \cos \varphi \, dr \, d\varphi$, where $c = 2\,\mathrm{length}(\partial \mathcal{D}) + 4\pi \mathbf{r}$ denotes the normalizing factor.

We recall a few standard facts from billiard theory [10, 11, 18, 96]. In the r, φ coordinates, u-curves are increasing and s-curves are decreasing, see (4.11) and (4.14), and they are uniformly transversal to each other. For every Q and integer m, the map $\mathcal{F}_Q^m$ is discontinuous on finite union of curves in Ω_0, which are stable for $m > 0$ and unstable for $m < 0$. The discontinuity curves of $\mathcal{F}_Q^m$ stretch continuously across Ω_0 between the two borders of Ω_0, i.e. from $\varphi = -\pi/2$ to $\varphi = \pi/2$ (they intersect each other, of course).

We use the $|\cdot|$-norm, see Section 4.2, to measure the lengths of stable and unstable curves. For a u-curve $W \subset \Omega_0$ and a point $x \in \Omega_0$ we define $\mathrm{dist}(x, W)$ to be the minimal length of s-curves connecting x with W, and vice versa (if there is no such connecting curve, we set $\mathrm{dist}(x, W) = \infty$). We define the "Hausdorff distance" between two u-curves $W_1, W_2 \subset \Omega_0$ to be

$$\mathrm{dist}(W_1, W_2) = \max\left\{ \sup_{x \in W_1} \mathrm{dist}(x, W_2), \sup_{y \in W_2} \mathrm{dist}(y, W_1) \right\}$$

(and similarly for s-curves). Let $W \subset \Omega_0$ be a stable or unstable curve with endpoints x_1 and x_2, and $\varepsilon < |W|/2$. For any two points $y_1, y_2 \in W$ such that $|W(x_i, y_i)| < \varepsilon$ for $i = 1, 2$, we call the middle segment $W(y_1, y_2)$ an ε-reduction of W.

Next we show that, in a certain crude sense, the map $\mathcal{F}_Q$ depends Lipschitz continuously on Q. Let Q, Q' satisfy (3.3) and $\varepsilon = \|Q - Q'\|$. The following lemma is a simple geometric observation:

LEMMA 4.11. *There are constants $c_2 > c_1 > 1$ such that*
 (a) *The discontinuity curves of the map $\mathcal{F}_{Q'}$ in Ω_0 are within the $c_1 \varepsilon$-distance of those of the map $\mathcal{F}_Q$.*
 (b) *Let $W \subset \Omega_0$ be a u-curve of length $> 2c_2 \varepsilon$ such that $\mathcal{F}_Q$ and $\mathcal{F}_{Q'}$ are smooth on W. Then there are two $c_2 \varepsilon$-reductions of this curve, $\tilde{W}$ and $\tilde{W}'$, such that $\mathrm{dist}(\mathcal{F}_Q(\tilde{W}), \mathcal{F}_{Q'}(\tilde{W}')) < c_1 \varepsilon$.*

COROLLARY 4.12. *There is a constant $c_3 > c_2$ such that for any H-curve $W \subset \Omega_0$ of length $> c_3 \varepsilon$ there are two finite partitions $W = \cup_{i=0}^{I} W_i = \cup_{i=0}^{I} W_i'$ of W such that*

(a) $|W_0| < c_3 \varepsilon$ and $|W_0'| < c_3 \varepsilon$
(b) *For each $i = 1, \ldots, I$, the sets $\mathcal{F}_Q(W_i)$ and $\mathcal{F}_{Q'}(W_i')$ are H-curves such that* $\mathrm{dist}(\mathcal{F}_Q(W_i), \mathcal{F}_{Q'}(W_i')) < c_1 \varepsilon$.

Proof. The singularities of $\mathcal{F}_Q$ divide W into $\leq K_1 = 1 + L_{\max}/L_{\min}$ pieces, see the proof of Lemma 4.10 (a), and so do the singularities of $\mathcal{F}_{Q'}$. Removing pieces shorter than $2c_2 \varepsilon$ and using Lemma 4.11 gives us two partitions $W = \cup_{j=0}^{J} \hat{W}_j = \cup_{j=0}^{J} \hat{W}_j'$ such that

(a) $|\hat{W}_0| < 2K_1 c_2 \varepsilon$ and $|\hat{W}_0'| < 2K_1 c_2 \varepsilon$
(b) For each $j = 1, \ldots, J$, the sets $\mathcal{F}_Q(\hat{W}_j)$ and $\mathcal{F}_{Q'}(\hat{W}_j')$ are u-curves such that $\mathrm{dist}(\mathcal{F}_Q(\hat{W}_j), \mathcal{F}_{Q'}(\hat{W}_j')) < c_1 \varepsilon$.

Next, for any homogeneity strip $\mathbb{H}_k$ and $j \geq 1$, consider the H-curves $\hat{W}_{jk} = \mathcal{F}_Q(\hat{W}_j) \cap \mathbb{H}_k$ and $\hat{W}_{jk}' = \mathcal{F}_{Q'}(\hat{W}_j') \cap \mathbb{H}_k$. It is easy to see that some $Cc_1\varepsilon$-reductions of these curves, call them W_{jk} and W_{jk}', respectively, are $c_1\varepsilon$-close to each other in the Hausdorff metric (here $C > 0$ is the bound on the slopes of u-curves and s-curves in Ω_0). Then we take the nonempty curves $\mathcal{F}_Q^{-1}(W_{jk})$ and $\mathcal{F}_{Q'}^{-1}(W_{jk}')$ for all j and k and relabel them to define the elements of our partitions W_i and W_i', respectively. Using the notation of Lemma 4.10 we have

$$|W \setminus \cup_i W_i| \leq 2K_1 c_2 \varepsilon + \big|\{x \in W : r_1(x) < Cc_1\varepsilon\}\big|$$
$$\leq 2K_1 c_2 \varepsilon + \beta_1 \vartheta^{-1} \big|\{x : r_0(x) < Cc_1 \vartheta \varepsilon\}\big| + \beta_2 \varepsilon |W|$$

(we apply the estimate in Lemma 4.10 (a) to $\mathcal{F}_Q$). The resulting bound clearly does not exceed $c_3 \varepsilon$ for some $c_3 > 0$. A similar bound holds for $|W \setminus \cup W_i'|$. □

COROLLARY 4.13. *There is a constant $c_4 > 1$ such that for each integer m the discontinuity sets of the maps $\mathcal{F}_Q^m$ and $\mathcal{F}_{Q'}^m$ are $c_4 \varepsilon$-close to each other in the Hausdorff metric.*

Proof. Let $c_4 = 10c_2/(1 - \vartheta)$. We prove this for $m > 0$ (the case $m < 0$ follows by time reversal) using induction on m. For $m = 1$ the statement follows from Lemma 4.11 (a). Assume that it holds for some $m \geq 1$. Now, if the statement fails for $m + 1$, then there is a point $x \in \Omega_0$ at which $\mathcal{F}_{Q'}^{m+1}$ is discontinuous and which lies in the middle of a u-curve W, $|W| = 2c_4 \varepsilon$, on which $\mathcal{F}_Q^{m+1}$ is smooth. We can assume that $\mathcal{F}_{Q'}$ is smooth on a $c_1 \varepsilon$-reduction $\hat{W}$ of W, too, otherwise we apply Lemma 4.11 (a). Now by Lemma 4.11 (b), there are $c_2 \varepsilon$-reductions of

$\hat{W}$, call them $\tilde{W}$ and $\tilde{W}'$, such that $\text{dist}(\mathcal{F}_Q(\tilde{W}), \mathcal{F}_{Q'}(\tilde{W}')) < c_1\varepsilon$. Due to our choice of ε_4 and the expansion of u-curves by a factor $\geq \vartheta^{-1}$, the point $\mathcal{F}_{Q'}(x)$ divides $\mathcal{F}_{Q'}(\tilde{W}')$ into two u-curves of length $> c_4\varepsilon + 5c_2\varepsilon$ each. Since $\mathcal{F}_{Q'}^m$ is discontinuous at $\mathcal{F}_{Q'}(x)$, our inductive assumption implies that $\mathcal{F}_Q^m$ is discontinuous on $\mathcal{F}_Q(\tilde{W})$, a contradiction. □

Let $Q \in \mathcal{D}$ satisfy (3.3) and $x \in \Omega$. Denote

(4.31) $$\varepsilon_n(x, Q) = \max_{0 \leq i \leq n} \|Q - Q(\mathcal{F}^i x)\| + 1/M$$

where $Q(y)$ denotes the Q-coordinate of a point $y \in \Omega$. For a u-curve $W \subset \Omega$ we put

$$\varepsilon_n(W, Q) = \sup_{x \in W} \varepsilon_n(x, Q)$$

Recall that the Q coordinate varies by $< \text{Const}/M$ on u-curves, so that the map $\mathcal{F}^n$ acts on a u-curve $W \subset \Omega$ similarly to the action of $\mathcal{F}_Q^n$ on its projection $\pi_Q(W)$, if $\varepsilon_n(W, Q)$ is small.

Recall that we consider initial conditions satisfying (3.10). In the lemmas below we require $n \leq \delta_\diamond \sqrt{M}$ to prevent collision of the heavy particle with the walls, see Section 3.3.

The following two lemmas are close analogies of the previous results (with, possibly, different values of the constants $c_1, \ldots, c_4$), and can be proved by similar arguments, so we omit details:

LEMMA 4.14. *There exist constants c_1, c_3 such that for any H-curve $W \subset \Omega$ of length $> c_3\varepsilon$ there are two finite partitions $W = \cup_{i=0}^I W_i = \cup_{i=0}^I W_i'$ of W such that*
  (a) $|W_0| < c_3\varepsilon$ and $|W_0'| < c_3\varepsilon$
  (b) *For each $i = 1, \ldots, I$, the sets $\mathcal{F}_Q(\pi_0(W_i))$ and $\pi_0(\mathcal{F}(W_i'))$ are H-curves such that $\text{dist}(\mathcal{F}_Q(\pi_0(W_i)), \pi_0(\mathcal{F}(W_i'))) < c_1\varepsilon$.*
Here $\varepsilon = \varepsilon_1(W, Q)$.

LEMMA 4.15. *There is a constant c_4 such that for any discontinuity point x of the map $\mathcal{F}^n$, $1 \leq n \leq \delta_\diamond\sqrt{M}$, its projection $\pi_0(x)$ lies in the $c_4\varepsilon$-neighborhood of some discontinuity curve of the map $\mathcal{F}_Q^n$, were $\varepsilon = \varepsilon_n(x, Q)$.*

LEMMA 4.16. *For any $1 \leq n \leq \delta_\diamond\sqrt{M}$ the singularity set $\mathcal{S}_n \subset \Omega$ of the map $\mathcal{F}^n$ is a finite union of smooth compact manifolds of codimension one with boundaries. For every $Q, V \in \Upsilon_{\delta_1}$ the manifold $\mathcal{S}_n$ intersects $\Omega_{Q,V}$ transversally (in fact, almost orthogonally), and $\mathcal{S}_n \cap \Omega_{Q,V}$ is a finite union of s-curves.*

Proof. The first claim follows from our finite horizon assumption.

Next, recall that $\mathcal{S}_n = \bigcup_{j=0}^{n-1} \mathcal{F}^{-j} \mathcal{S}_0$. Consider for example a component $\hat{\mathcal{S}} \subset \mathcal{S}_0$ corresponding to a grazing collision between particles (other components can be treated similarly). In the whole 8-dimensional phase space $\hat{\mathcal{S}}$ is given by the equations

(4.32) $$\|Q - q\| = \mathbf{r} \qquad \text{(collision)}$$
(4.33) $$\langle Q - q, V - v \rangle = 0 \qquad \text{(tangency)}$$
(4.34) $$MV^2 + v^2 = 1 \qquad \text{(energy conservation)}$$

Recall that $\mathcal{F}$ preserves the restriction to Ω of the symplectic form

$$\omega\big((dQ_1, dV_1, dq_1, dv_1), (dQ_2, dV_2, dq_2, dv_2)\big)$$
$$= M \langle dQ_1, dV_2 \rangle - M \langle dQ_2, dV_1 \rangle + \langle dq_1, dv_2 \rangle - \langle dq_2, dv_1 \rangle$$

By (4.32)–(4.34) the tangent space $\mathcal{T}\hat{\mathcal{S}}$ in the whole 8-dimensional space is the skew-orthogonal complement of the linear subspace spanning three vectors

$$\begin{array}{rcl}
e_1 &=& \left(\; 0, \quad \frac{q-Q}{M}, \quad 0, \quad Q-q \;\right) \\
e_2 &=& \left(\; \frac{Q-q}{M}, \quad \frac{v-V}{M}, \quad q-Q, \quad V-v \;\right) \\
e_3 &=& \left(\; V, \quad 0, \quad v, \quad 0 \;\right)
\end{array}$$

Equivalently, in $\mathcal{T}\Omega$, the subspace $\mathcal{T}\hat{\mathcal{S}}$ can be described as the skew-orthogonal complement of

$$\mathrm{span}(e_1, e_2, e_3) \bigcap \mathcal{T}\Omega = \mathbb{R}e_1,$$

where $\mathbb{R}e_1 = \{ce_1, c \in \mathbb{R}\}$. Observe that e_1 is tangent to Ω because

$$\omega(e_1, e_3) = \langle Q - q, v - V \rangle = 0$$

by (4.33). Hence $\mathcal{T}(\mathcal{F}^{-j}\hat{\mathcal{S}})$ is the skew-orthogonal complement of $D\mathcal{F}^{-j}(e_1)$ and so $\mathcal{T}\big((\mathcal{F}^{-j}\hat{\mathcal{S}}) \cap \Omega_{Q,V}\big)$ is the skew-orthogonal complement of $\pi\big(D\mathcal{F}^{-j}(e_1)\big)$. Since $(\mathcal{F}^{-j}\hat{\mathcal{S}}) \cap \Omega_{Q,V}$ is one-dimensional,

$$\mathcal{T}\big((\mathcal{F}^{-j}\hat{\mathcal{S}}) \cap \Omega_{Q,V}\big) = \mathbb{R}\,\pi\big(D\mathcal{F}^{-j}(e_1)\big).$$

Our results in Section 4.1 easily imply that $\pi\big(D\mathcal{F}^{-j}(e_1)\big)$ is an s-vector for all $j \geq 1$, so the lemma follows. $\square$

Now let A be a function from Proposition 3.3. For each pair (Q, V) we define a function $A_{Q,V}$ on Ω_0 by

(4.35) $$A_{Q,V} = A \circ (\pi_0|_{\Omega_{Q,V}})^{-1}$$

Note that $A_{Q,V}$ has discontinuities on the set $\mathcal{S}_{Q,V} = \pi_0(\mathcal{S}_{n_A} \cap \Omega_{Q,V})$. Also,
$$\bar{A}(Q,V) = \int_{\Omega_{Q,V}} A(Q,V,q,w)\, d\mu_{Q,V}(q,w) = \int_{\Omega_0} A_{Q,V}(r,\varphi)\, d\mu_0$$

LEMMA 4.17. *For any (Q,V) and (Q',V') we have*
$$|\bar{A}(Q,V) - \bar{A}(Q',V')| \leq C \Big(\|Q-Q'\| + \|V-V'\|$$
$$+ n_A(\|V\| + \|V'\|) + n_A^2/M\Big)$$
for some $C = C(A) > 0$.

Proof. If A had a bounded local Lipschitz constant (3.9) on the entire space Ω, the estimate would be trivial. However, the function A is allowed to have singularities on $\mathcal{S}_{n_A}$ (the discontinuity set for the map $\mathcal{F}^{n_A}$), and the local Lipschitz constant $\mathrm{Lip}_x A$ is allowed to grow near $\mathcal{S}_{n_A}$, according to Lemma 3.1. As a result, two error terms appear, denoted by $E_1 + E_2$, where E_1 comes from the fact that the functions $A_{Q,V}$ and $A_{Q',V'}$ have different singularity sets $\mathcal{S}_{Q,V}$ and $\mathcal{S}_{Q',V'}$, and E_2 comes from the growing local Lipschitz constant near these singularity sets.

The error term E_1 is bounded by $2\|A\|_\infty \mathrm{Area}(G)$, where G is the region swept by the singularity set $\mathcal{S}_{Q,V}$ as it transforms to $\mathcal{S}_{Q',V'}$ when (Q,V) continuously change to (Q',V'). According to Lemma 4.15, these singularity sets lie within the $c_4\varepsilon'$-distance from the discontinuity lines of the maps $\mathcal{F}_Q^{n_A}$ and $\mathcal{F}_{Q'}^{n_A}$, respectively, where
$$\varepsilon' = \mathrm{Const}\,[\,n_A(\|V\| + \|V'\|) + n_A^2/M\,]$$
Due to our finite horizon assumption, the discontinuity lines of $\mathcal{F}_Q^{n_A}$ and $\mathcal{F}_{Q'}^{n_A}$ have a finite total length, and they lie within the $c_4\|Q-Q'\|$-distance from each other by Corollary 4.13. Therefore, we can cover G by a finite union of stripes of width $2c_4\varepsilon' + c_4\|Q-Q'\|$ bounded by s-curves roughly parallel to the discontinuity lines of $\mathcal{F}_Q^{n_A}$ and $\mathcal{F}_{Q'}^{n_A}$. The union of these stripes, call it G_0, has area bounded by $C(\varepsilon' + \|Q-Q'\|)$, where $C = C(n_A) > 0$, thus
$$\left|\int_{G_0} [A_{Q,V}(r,\varphi) - A_{Q',V'}(r,\varphi)]\, d\mu_0\right| \leq 2\|A\|_\infty C(\varepsilon' + \|Q-Q'\|)$$
To estimate the error term E_2, we note that the Lipschitz constants of the functions $A_{Q,V}$ and $A_{Q',V'}$ on the domain $\Omega_0 \setminus G_0$ are bounded by $CL_A \mathrm{dist}(x,G_0)^{-\beta_A}$, where $x = (r,\varphi) \in \Omega_0$. Here the distance, originally measured in the Lebesgue metric in Lemma 3.1, can also be

measured in the equivalent $|\cdot|$ metric introduced above (the length of the the shortest u-curve connecting x with ∂G_0). Thus,

$$E_2 \leq CL_A \left(\|Q - Q'\| + \|V - V'\| \right) \int_{\Omega_0 \setminus G_0} [\operatorname{dist}(x, G_0)]^{-\beta_A} \, d\mu_0$$

and the integral here is finite because $\beta_A < 1$. □

4.6. Equidistribution properties. We use the following scheme to estimate $\mathbb{E}_\ell(A \circ \mathcal{F}^n)$, where $\ell = (\gamma, \rho)$ is a standard pair in Proposition 3.3.

Let n_1, n_2 (to be chosen later) satisfy $K\big|\ln|\gamma|\big| < n_1 < n_2 < n$. For each point $x \in \gamma$ put

$$k(x) = \min_{n_1 < k < n_2} \{ k \colon |\gamma_k(x)| \geq \varepsilon_0 \}$$

where $\gamma_k(x)$ denotes the H-component of $\mathcal{F}^k(\gamma)$ that contains the point $\mathcal{F}^k(x)$ (the constant ε_0 was introduced in Lemma 4.10). In other words, $k(x)$ is the first time, during the time interval (n_1, n_2), when the image of the point x belongs in an H-curve of length $\geq \varepsilon_0$. Clearly, the set $\{\mathcal{F}^{k(x)}(x) \colon x \in \gamma\}$ is a union of H-curves of length $> \varepsilon_0$. We denote those curves by γ_j, $j \geq 1$, and for each γ_j denote by $k_j \in (n_1, n_2)$ the iteration of $\mathcal{F}$ at which this curve was created. Let ρ_j be the density of the measure $\mathcal{F}^{k_j}(\mathrm{mes}_\ell)$ conditioned on γ_j. Observe that (γ_j, ρ_j) is a standard pair for every $j \geq 1$.

The function $k(x)$ may not be defined on some parts of γ, but by Lemma 4.10 we have

$$\mathrm{mes}_\ell \big(x \in \gamma \colon k(x) \text{ is not defined} \big) \leq Cq^{n_2 - n_1}$$

Hence

(4.36) $$\mathbb{E}_\ell(A \circ \mathcal{F}^n) = \sum_j c_j \mathbb{E}_{\ell_j}(A \circ \mathcal{F}^{n-k_j}) + \mathcal{O}(q^{n_2 - n_1})$$

where $\sum_j c_j > 1 - Cq^{n_2 - n_1}$.

We now analyze each standard pair (γ_j, ρ_j) separately, and we drop the index j for brevity. For example, we denote by $\mathrm{mes}_\ell = \mathrm{mes}_{\ell_j}$ the measure on $\gamma = \gamma_j$ with density $\rho = \rho_j$. Let $x \in \gamma$ be an arbitrary point and $(Q, V) = \pi_1(x)$ its coordinates. For each $0 \leq i \leq n - k$ consider the map

(4.37) $$\mathcal{F}_i \colon = \mathcal{F}_Q^{n-k-i} \circ \pi_0 \circ \mathcal{F}^i$$

on the curve γ (here, as in the previous section, we identify the domain of the map $\mathcal{F}_Q$ with Ω_0). Note that $\mathcal{F}_{n-k} = \pi_0 \circ \mathcal{F}^{n-k}$, and so $A \circ \mathcal{F}^{n-k} =$

$A_{Q_{n-k},V_{n-k}} \circ \mathcal{F}_{n-k}$, where the function $A_{Q,V}$ on Ω_0 was defined by (4.35). Our further analysis is based on the obvious identity

$$\mathbb{E}_\ell(A \circ \mathcal{F}^{n-k}) - \bar{A}(Q,V) = \mathbb{E}_\ell(A_{Q,V} \circ \mathcal{F}_0) - \bar{A}(Q,V)$$
$$(4.38) \qquad\qquad + \sum_{i=0}^{n-k-1} \left[\mathbb{E}_\ell(A_{Q,V} \circ \mathcal{F}_{i+1}) - \mathbb{E}_\ell(A_{Q,V} \circ \mathcal{F}_i) \right]$$
$$+ \mathbb{E}_\ell\left((A_{Q_{n-k},V_{n-k}} \circ \mathcal{F}_{n-k}) - (A_{Q,V} \circ \mathcal{F}_{n-k}) \right)$$

We divide the estimate of (4.38) into three parts (Propositions 4.18, 4.19 and 4.21).

PROPOSITION 4.18. *We have*
$$\left| \mathbb{E}_\ell(A_{Q,V} \circ \mathcal{F}_0) - \bar{A}(Q,V) \right| \leq C\theta_0^{n-k}$$
for some constants $C > 0$ and $\theta_0 < 1$.

Proof. Since $\mathcal{F}_0 = \mathcal{F}_Q^{n-k} \circ \pi_0$, our proposition asserts the equidistribution for dispersing billiards, see Appendix A.1. Note that the u-curve γ has length of order one ($|\gamma| > \varepsilon_0$), hence there is no "waiting period" during which the curve needs to expand – the exponential convergence starts right away. Furthermore, the convergence is uniform in Q, i.e. C and θ_0 are independent of Q and V, see Extension 1 in Appendix A. □

PROPOSITION 4.19. *For each $0 \leq i \leq n - k - 1$ we have*
$$(4.39) \qquad \left| \mathbb{E}_\ell(A_{Q,V} \circ \mathcal{F}_{i+1}) - \mathbb{E}_\ell(A_{Q,V} \circ \mathcal{F}_i) \right| \leq C\varepsilon_\gamma$$
where $C > 0$ is a constant and
$$\varepsilon_\gamma := (n-k)\|V\| + (n-k)^2/M$$
$$\geq c \max_{0 \leq j \leq n-k} \sup_{x \in \gamma} \left(\|Q - Q(\mathcal{F}^j x)\| + \|V - V(\mathcal{F}^j x)\| \right)$$
where $c > 0$ is a small constant.

Proof. Estimates of this kind have been obtained for Anosov diffeomorphisms [34] and they are based on shadowing type arguments. We follow this line of arguments here, too, but face additional problems when dealing with singularities.

We first outline our proof. We will construct two subsets $\gamma_i^*, \hat{\gamma}_i^* \subset \gamma$ and an absolutely continuous map $H^* \colon \gamma_i^* \to \hat{\gamma}_i^*$ (in fact, the map $h^* = \mathcal{F}_{i+1} \circ H^* \circ \mathcal{F}_i^{-1}$ will be the holonomy map between some H-components of $\mathcal{F}_i(\gamma)$ and those of $\mathcal{F}_{i+1}(\gamma)$) that have three properties:

(H1) $\mathrm{mes}_\ell(\gamma \setminus \gamma_i^*) < C\varepsilon_\gamma$ and $\mathrm{mes}_\ell(\gamma \setminus \hat{\gamma}_i^*) < C\varepsilon_\gamma$,
(H2) $\mathbb{E}_{\ell_i^*}(|A_{Q,V} \circ \mathcal{F}_i - A_{Q,V} \circ h^* \circ \mathcal{F}_i|) < C\varepsilon_\gamma \vartheta^{n-k-i}$,

4. STANDARD PAIRS

(H3) the Jacobian $\mathcal{J}_*(x)$ of the map H^* satisfies

(4.40)
$$|\ln \mathcal{J}_*| \leq 2$$

and

(4.41)
$$\mathbb{E}_{\ell_i^*}(|\ln \mathcal{J}_*|) < C\varepsilon_\gamma$$

(here $\mathbb{E}_{\ell_i^*}(f) = \int_{\gamma_i^*} f \, d\mathrm{mes}_\ell$ for any function f). In the rest of the proof of Proposition 4.19, we will denote $A_{Q,V}$ by A for brevity.

Observe that (H1)–(H3) imply (4.39). Indeed, we use the change of variables $y = H^*(x)$ and get

(4.42)
$$\left| \mathbb{E}_\ell(A \circ \mathcal{F}_{i+1} - A \circ \mathcal{F}_i) \right|$$

$$\leq \left(\int_{\gamma - \gamma_i^*} |(A \circ \mathcal{F}_i) \, d\mathrm{mes}_{l_i}| + \int_{\gamma - \hat{\gamma}_i^*} |(A \circ \mathcal{F}_{i+1})| \, d\mathrm{mes}_{l_i} \right)$$
$$+ \mathbb{E}_{l_i^*} |(A \circ h^* \circ \mathcal{F}_i) - (A \circ \mathcal{F}_i)|$$
$$+ \mathbb{E}_{l_i^*} |(A \circ h^* \circ \mathcal{F}_i)(\mathcal{J}_* - 1)|$$
$$= I + II + III$$

Next

(4.43) $\quad |I| \leq 2C\varepsilon_\gamma \|A\|_\infty \quad\quad\quad$ by (H1)

(4.44) $\quad |II| \leq C\varepsilon_\gamma \quad\quad\quad$ by (H2)

To estimate III observe that (4.40) implies $|\mathcal{J}_* - 1| \leq \mathrm{Const}\,|\ln \mathcal{J}_*|$ on γ_i^*, so

(4.45) $\quad |III| \leq C \mathbb{E}_{\ell_i^*}(|\ln \mathcal{J}_*|) \|A\|_\infty \leq C\varepsilon_\gamma \|A\|_\infty \quad$ by (H3)

This completes the proof of (4.39) assuming (H1)–(H3).

We begin the construction of the sets γ_i^* and $\hat{\gamma}_i^*$. First, the definition of both maps $\mathcal{F}_i$ and $\mathcal{F}_{i+1}$, see (4.37), involves the transformation of the curve γ to $\mathcal{F}^i(\gamma)$. Let $\tilde{\gamma}$ be an H-component of $\mathcal{F}^i(\gamma)$. If its length is $< c_3 \varepsilon_\gamma$, we simply discard it (i.e., remove its preimage in γ from the construction of both γ_i^* and $\hat{\gamma}_i^*$). If $|\tilde{\gamma}| > c_3 \varepsilon_\gamma$, then Lemma 4.14 gives us two partitions $\tilde{\gamma} = \cup_{j=0}^J \tilde{\gamma}_i = \cup_{j=0}^J \tilde{\gamma}_j'$, such that for each $j = 1, \ldots, J$ the sets $\mathcal{F}_Q(\pi_0(\tilde{\gamma}_j))$ and $\pi_0(\mathcal{F}(\tilde{\gamma}_j'))$ are H-curves and $\mathrm{dist}(\mathcal{F}_Q(\pi_0(\tilde{\gamma}_j)), \pi_0(\mathcal{F}(\tilde{\gamma}_j'))) < c_1 \varepsilon_\gamma$. We remove the preimage of $\tilde{\gamma}_0$ from the construction of γ_i^*, and the preimage of $\tilde{\gamma}_0'$ from the construction of $\hat{\gamma}_i^*$. By Lemma 4.10, the total mes_ℓ-measure of the (so far) removed sets is $\mathcal{O}(\varepsilon_\gamma)$. The remaining H-curves in $\mathcal{F}_Q(\pi_0(\mathcal{F}^i(\gamma)))$ and $\pi_0(\mathcal{F}^{i+1}(\gamma))$ are now paired according to Lemma 4.14.

Consider an arbitrary pair of curves $W' \subset \mathcal{F}_Q(\pi_0(\mathcal{F}^i(\gamma)))$ and $W'' \subset \pi_0(\mathcal{F}^{i+1}(\gamma))$ constructed above and remember that $\mathrm{dist}(W', W'') <$

$c_1\varepsilon_\gamma$. According to our definition of the maps $\mathcal{F}_i$ and $\mathcal{F}_{i+1}$, both curves W' and W'' will be then iterated $n-k-i-1$ times under the same billiard map $\mathcal{F}_Q$. For each $x \in W'$ and $n \geq 0$ denote by $r_n(x)$ the distance from the point $\mathcal{F}_Q^n(x)$ to the nearest endpoint of the H-component of $\mathcal{F}_Q^n(\gamma')$ that contains the point $\mathcal{F}_Q^n(x)$. Define

$$(4.46) \qquad W'_* = \{x \in W' : \ r_n(x) \geq C\varepsilon_\gamma \vartheta^n \ \text{ for all } \ n \geq 0\}$$

where C is a constant chosen as follows. Let $r^s(x)$ denote the distance from x to the nearest endpoint of the homogeneous stable manifold W_x^s for the map $\mathcal{F}_Q$ passing through x. (A homogeneous stable manifold $W^s \subset \Omega_Q$ is a maximal curve such that $\mathcal{F}_Q^n(W^s)$ is a homogeneous s-curve for each $n \geq 0$.) By [11, Appendix 2], if $r^s(x) < \varepsilon$, then for some $n \geq 0$ the point $\mathcal{F}_Q^n(x)$ lies within the $(\varepsilon\vartheta^n)$-neighborhood of either a singularity set of the map $\mathcal{F}_Q$ or the boundary of a homogeneity strip $\mathbb{H}_{\pm k}$, $k \geq k_0$. Since the singularity lines and the boundaries of homogeneity strips are uniformly transversal to u-curves it follows that if C in (4.46) is large enough then for all $x \in W'_*$ $W_x^s \cap W'' \neq \emptyset$. Let $h\colon W'_* \to W''$ denote the holonomy map (defined by sliding along the stable manifolds W_x^s). We remove the preimage of the set $W' \setminus W'_*$ from the construction of γ_i^*, and the preimage of the set $W'' \setminus h(W'_*)$ – from the construction of $\hat\gamma_i^*$.

We need to estimate the measure of the sets just removed from the construction. Denote by $\gamma' = \cup_\alpha \gamma'_\alpha \subset \Omega_0$ the union of the above H-curves $W' \subset \mathcal{F}_Q(\pi_0(\mathcal{F}^i(\gamma)))$ and by $\text{mes}_{\gamma'}$ the restriction of the measure $\mathcal{F}_Q(\pi_0(\mathcal{F}^i(\text{mes}_\ell)))$ to γ'.

CLAIM. $\text{mes}_{\gamma'}\bigl(\cup_{W'}(W' \setminus W'_*)\bigr) \leq \text{Const}\,\varepsilon_\gamma$, and a similar estimate holds for $\cup_{W''}(W'' \setminus h(W'_*))$.

Proof. For any $n \geq 0$ and $\varepsilon > 0$

$$(4.47) \qquad \text{mes}_{\gamma'}\left(x \in \gamma' : r_n(x) < \varepsilon\right) < \beta\varepsilon$$

where $\beta > 0$ is some large constant, according to the remarks in the end of Section 4.3 (they are stated for the map $\mathcal{F}$, but obviously apply to the billiard map $\mathcal{F}_Q$ as well).

Thus

$$\text{mes}_{\gamma'}\bigl(\cup_{W'}(W' \setminus W'_*)\bigr) \leq \sum_{n=0}^\infty C\beta\varepsilon_\gamma\vartheta^n = \frac{C\beta}{1-\vartheta}\varepsilon_\gamma.$$

This proves the estimate for $\text{mes}_{\gamma'}\bigl(\cup_{W'}(W'\setminus W'_*)\bigr)$. To get a similar estimate for $\cup_{W''}(W'' \setminus h(W'_*))$ we observe that the $\mathcal{F}_Q$ orbits of the points $x \in \cup_{W''}(W'' \setminus h(W'_*))$ also come close to the singularities. Indeed, if $r^s(x) \leq \text{Const}\,\varepsilon_\gamma$, then the orbit of x comes close to singularities by the previous discussion. If the opposite inequality holds, then the orbit of x should pass near a singularity since otherwise we would have $h^{-1}(x) \in W'_*$. Now the result follows by (4.47). $\square$

This completes the construction of the sets γ_i^* and $\hat{\gamma}_i^*$ and the proof of (H1). The map $h^*: \mathcal{F}_i(\gamma_i^*) \to \mathcal{F}_{i+1}(\hat{\gamma}_i^*)$ is the induced holonomy map. It remains to prove (H2) and (H3).

Put $d := n - k - i - 1$ for brevity. For any point $x' \in W'_*$ and its "sister" $x'' = h(x) \in W''$, the points $z' = \mathcal{F}_Q^d(x') \in \mathcal{F}_i(\gamma)$ and $z'' = \mathcal{F}_Q^d(x'') \in \mathcal{F}_{i+1}(\gamma)$ (related by $h^*(z') = z''$) will be $(C\vartheta^d \varepsilon_\gamma)$-close, since $\mathcal{F}_Q$ contracts stable manifolds by a factor $\leq \vartheta < 1$. In other words, the trajectory of the point x'' shadows that of x' in the forward dynamics. Therefore, the values of the function $A = A_{Q,V}$ will differ at the endpoints z' and z'' by at most $\mathcal{O}(\vartheta^d \varepsilon_\gamma D(z', z''))$, unless they are separated by a discontinuity curve of the function A. Here $D(z', z'') = [\text{dist}(W^s(z', z''), \mathcal{S}_{Q,V})]^{-\beta_A}$, where $W^s(z', z'')$ denotes the stable manifold connecting z' with z'', and $\mathcal{S}_{Q,V} = \pi_0(\mathcal{S}_{n_A} \cap \Omega_{Q,V})$ in accordance with Lemma 3.1 (b).

Let W° be an H-component of $\mathcal{F}_Q^d(W')$. Put $W_*^\circ = W^\circ \cap \mathcal{F}_Q^d(W'_*)$ and $\text{mes}_i = \mathcal{F}_i(\text{mes}_\ell)$. We need to estimate

$$\Delta(W^\circ) := \int_{W_*^\circ} |A(z') - A(z'')| \, d\,\text{mes}_i$$

The curve W° crosses the discontinuity set $\mathcal{S}_{Q,V}$ in at most K_{n_A} points, cf. Lemma 4.16. If a pair of points z' and $z'' = h^*(z')$ is separated by a curve of $\mathcal{S}_{Q,V}$, then both z' and z'' lie in the $(C\vartheta^d \varepsilon_\gamma)$-neighborhood of that curve. Let $\mathfrak{U}_A$ denote the $(C\vartheta^d \varepsilon_\gamma)$-neighborhood of $\mathcal{S}_{Q,V}$. The sets $W^\circ \cap \mathfrak{U}_A$ and $W_*^\circ \cap \mathfrak{U}_A$ have $|\cdot|$-measure less than K_{n_A} times the $|\cdot|$ measure of the $(C\vartheta^d \varepsilon_\gamma)$-neighborhood of the endpoints of W° and W_*°, respectively. Hence the contribution of these sets to $\Delta(W^\circ)$ will be bounded by

$$K_{n_A} \text{mes}_\ell \big(r_{n-k}(x) < C\vartheta^d \varepsilon_\gamma\big) \leq \text{Const}\, \vartheta^d \varepsilon_\gamma$$

where we used Lemma 4.10.

Next, the set $W^\circ \setminus \mathfrak{U}_A$ is a union of H-curves $W_1, \ldots, W_k$ with some $k \leq K_{n_A}$. For each W_j we put $W_{j*} = W_j \cap \mathcal{F}_Q^d(W'_*)$ and estimate

$$\int_{W_{j*}} |A(z') - A(z'')| \, d\,\text{mes}_i \leq C' \vartheta^d \varepsilon_\gamma \frac{\text{mes}_i(W_j)}{|W_j|} \int_0^{|W_j|} t^{-\beta_A} \, dt$$

$$\leq C'' \vartheta^d \varepsilon_\gamma \frac{\text{mes}_i(W_j)}{|W_j|^{\beta_A}}$$

where $C', C'' > 0$ are some constants. Summing up over j gives

$$(4.48) \quad \int_{W^\circ} |A(z') - A(z'')| \, d\,\mathrm{mes}_i \leq \mathrm{Const}\, K_{n_A}^{\beta_A} \vartheta^d \varepsilon_\gamma \frac{\mathrm{mes}_i(W^\circ)}{|W^\circ|^{\beta_A}},$$

where we first used the homogeneity of the measure mes_i to estimate

$$\mathrm{mes}_i(W_j) \leq \mathrm{Const}\, |W_j| \frac{\mathrm{mes}_i(W^\circ)}{|W^\circ|}$$

and then by Jensen's inequality obtain

$$\sum_j |W_j|^{1-\beta_A} \leq K_{n_A}^{\beta_A} |W^\circ|^{1-\beta_A}.$$

Next, summing over all the H-components of $\mathcal{F}_Q^d(W')$ and all the curves $W' \subset \mathcal{F}_Q(\pi_0(\mathcal{F}^i(\gamma)))$ gives a bound

$$\mathbb{E}_{\ell_i^*}(|A \circ \mathcal{F}_i - A \circ h^* \circ \mathcal{F}_i|) \leq \sum_{W^\circ \subset \mathcal{F}_i(\gamma)} \mathrm{Const}\, K_{n_A}^{\beta_A} \vartheta^d \varepsilon_\gamma \frac{\mathrm{mes}_i(W^\circ)}{|W^\circ|^{\beta_A}}$$

$$(4.49) \qquad\qquad \leq \mathrm{Const}\, K_{n_A}^{\beta_A} \vartheta^d \varepsilon_\gamma$$

where the last inequality follows from Lemma 4.10 and

$$(4.50) \quad \sum_{W^\circ \subset \mathcal{F}_i(\gamma)} \frac{\mathrm{mes}_i(W^\circ)}{|W^\circ|^{\beta_A}} \leq \mathrm{Const} \int_\gamma [r_{n-k}(x)]^{-\beta_A} \, d\rho(x) \leq \mathrm{Const}$$

(we remind the reader that $\beta_A < 1$). This proves (H2). It remains to prove (H3).

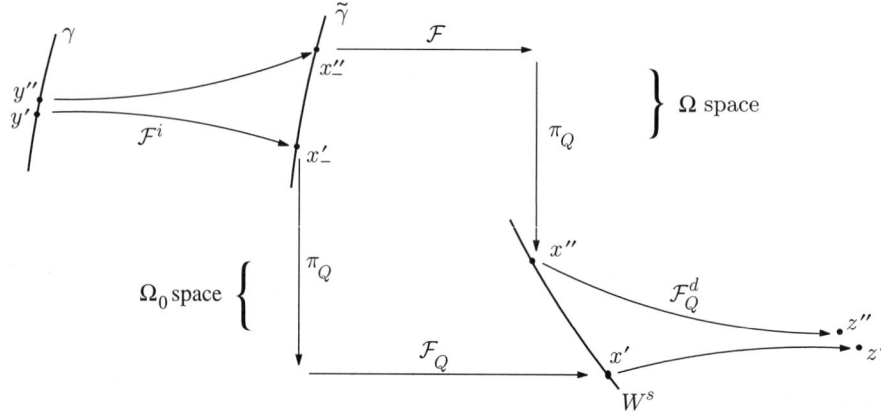

FIGURE 2. The construction in Proposition 4.19

Let $x'_-, x''_- \in \tilde{\gamma}$ be the preimages of x', x'', respectively, i.e. $x' = \mathcal{F}_Q(\pi_0(x'_-))$ and $x'' = \pi_0(\mathcal{F}(x''_-))$. Note that the distance between x'_- and x''_- is $< C\varepsilon_\gamma$. Let $y' = \mathcal{F}^{-i}(x'_-)$ and $y'' = \mathcal{F}^{-i}(x''_-)$ be the preimages of our two points on the original curve γ. Note that $\mathrm{dist}(y', y'') \leq C\vartheta^i \varepsilon_\gamma$ and $z' = \mathcal{F}_i(y')$ and $z'' = \mathcal{F}_{i+1}(y'')$, see Fig. 2. (We can say that the trajectory of the point y'' shadows that of y' during all the $n - k$ iterations.) Now $y'' = H^*(y')$, where $H^* = \mathcal{F}_{i+1}^{-1} \circ h^* \circ \mathcal{F}_i$. The Jacobian $\mathcal{J}_*$ of the map $H^* \colon \gamma \to \gamma$ satisfies

$$\ln \mathcal{J}_*(y') = \ln \frac{\mathcal{J}_\gamma \mathcal{F}_i(y')}{\mathcal{J}_\gamma \mathcal{F}_{i+1}(y'')} + \ln \mathcal{J} h^*(z')$$

where $\mathcal{J}_\gamma \mathcal{F}_i$ and $\mathcal{J}_\gamma \mathcal{F}_{i+1}$ denote the Jacobians (the expansion factors) of the maps $\mathcal{F}_i$ and $\mathcal{F}_{i+1}$, respectively, restricted to the curve γ, and $\mathcal{J} h^*$ is the Jacobian of the holonomy map h^*.

LEMMA 4.20. *We have*

$$(4.51) \qquad \left| \ln \frac{\mathcal{J}_\gamma \mathcal{F}_i(y')}{\mathcal{J}_\gamma \mathcal{F}_{i+1}(y'')} \right| \leq \frac{C\varepsilon_\gamma}{|\tilde{\gamma}|^{2/3}} + \sum_{r=0}^{d} \frac{C\vartheta^r \varepsilon_\gamma}{|\tilde{\gamma}'_r|^{2/3}}$$

and

$$(4.52) \qquad \left| \ln \mathcal{J} h^*(z') \right| \leq \sum_{r=d}^{\infty} \frac{C\vartheta^r \varepsilon_\gamma}{|\tilde{\gamma}'_r|^{2/3}}$$

where $\tilde{\gamma}'_r$ denotes the H-component of $\mathcal{F}_Q^{r+1}(\pi_0(\tilde{\gamma}))$ containing the point $\mathcal{F}_Q^r(x')$.

This lemma is, in a sense, an extension of Proposition 4.6. Its proof is given in Appendix C, after the proof of Proposition 4.6.

Now (4.40) follows directly from Lemma 4.20 and the definition of γ_i^*, cf. (4.46). To complete the proof of (H3), we need to establish (4.41):

$$\mathbb{E}_{\ell_i^*}(|\ln \mathcal{J}_*|) \leq \sum_{\tilde{\gamma} \subset \mathcal{F}^i(\gamma)} C\varepsilon_\gamma \frac{\mathrm{mes}'(\tilde{\gamma})}{|\tilde{\gamma}|^{2/3}}$$
$$+ \sum_{r=0}^{\infty} \sum_{\tilde{\gamma}' \subset \mathcal{F}_Q^{r+1}(\pi_0(\mathcal{F}^i(\gamma)))} C\vartheta^r \varepsilon_\gamma \frac{\mathrm{mes}'_r(\tilde{\gamma}')}{|\tilde{\gamma}'|^{2/3}}$$
$$\leq C\varepsilon_\gamma + \sum_{r=0}^{\infty} C\vartheta^r \varepsilon_\gamma = \mathrm{Const}\,\varepsilon_\gamma$$

where $\mathrm{mes}' = \mathcal{F}^i(\mathrm{mes}_\ell)$ and $\mathrm{mes}'_r = \mathcal{F}_Q^{r+1}(\pi_0(\mathcal{F}^i(\mathrm{mes}_\ell)))$. Here we used the same trick as in (4.50). The property (H3) is proved, and so is Proposition 4.19. □

PROPOSITION 4.21. *There is a constant C such that*
$$\left|\mathbb{E}_\ell\left((A_{Q_{n-k},V_{n-k}} \circ \mathcal{F}_{n-k}) - (A_{Q,V} \circ \mathcal{F}_{n-k})\right)\right| \leq C\varepsilon_\gamma.$$

Proof. The proof of this proposition follows exactly the same arguments as the proof of (4.49) in the estimate of (H2) so we omit it. □

We now return to our main identity (4.38) and obtain
$$\mathbb{E}_\ell(A \circ \mathcal{F}^{n-k}) - \bar{A}(Q,V) \leq C\theta_0^{n-k} + C(n-k)\varepsilon_\gamma$$
$$\leq C\theta_0^{n-k} + C(n-k)^2\|V\| + C(n-k)^3/M$$

Equation (4.36) now yields
$$\mathbb{E}_\ell(A \circ \mathcal{F}^n) = \sum_j c_j \bar{A}(Q_j, V_j) + \mathcal{O}(q^{n_2-n_1})$$
$$+ \mathcal{O}(\theta_0^{n-n_2}) + \mathcal{O}((n-n_1)^2) \max_{\gamma_j} \|V_j\|$$
$$+ \mathcal{O}\left((n-n_1)^3\right)/M$$
where $(Q_j, V_j) \in \pi_1(\gamma_j)$. Note that
$$\max_j \|V_j\| \leq \|V\| + Cn^2/M$$

Finally, we apply Lemma 4.17 to estimate the value of $\bar{A}(Q_j, V_j)$:
$$\left|\bar{A}(Q_j, V_j) - \bar{A}(Q,V)\right| \leq Cn\|V\| + Cn^2/M$$
and arrive at
$$\left|\mathbb{E}_\ell(A \circ \mathcal{F}^n) - \bar{A}(Q,V)\right| \leq C\|V\|\left[n + (n-n_1)^2\right]$$
$$+ C\left[n^2 + (n-n_1)^3\right]/M$$
$$+ C\,q^{n_2-n_1} + C\,\theta_0^{n-n_2}$$

Now for any $m \leq \min\{n/2, K \ln M\}$ we can choose n_1, n_2 so that $n - n_2 = n_2 - n_1 = m$. This completes the proof of Proposition 3.3. □

Remark. Proposition 3.3 can be easily generalized to any finite or countable union $\ell = \cup_\alpha \gamma_\alpha$ of H-curves with a smooth probability measure mes_ℓ on it (as introduced in the end of Section 4.4) provided (i) they have approximately the same Q and V coordinates and (ii) the lower bound on n (that is, $n \geq K|\ln \text{length}(\gamma)|$) is modified accordingly. For the latter, let us assume that
$$\text{mes}_\ell\left(r_0(x) < \varepsilon\right) < \beta_7 \varepsilon$$

for all $\varepsilon > 0$, in the notation of the end of Section 4.4; i.e. the curves γ_α are 'long' on the average. Then the proposition would simply hold for all $n \geq \text{Const}$.

To prove Corollary 3.4, we decompose the set $\mathcal{F}^{n-j}(\gamma)$ into H-components according to Proposition 3.2, then apply Proposition 3.3 to these H-components (see the above remark), and deal only with the last j iterations of $\mathcal{F}$. □

Remark. As it was mentioned in Chapter 3, the estimates of Proposition 3.3 and Corollary 3.4 are apparently less than optimal. Even though they suffice for our purposes, we outline possible improvements of the key estimates (4.38) and (4.42) used in our analysis. Such improvements could be useful in other applications.

- In our estimation of the term I in (4.42), there is no need to discard the orbits which pass close to singularities for small i (i.e., for $n - i \geq \text{Const} \ln \varepsilon_\gamma$), because the corresponding pieces expand in the remaining time and their images become asymptotically equidistributed.
- In the same estimation, if i is large then one can expect that the singularities of $\mathcal{F}$ on the set $\mathcal{F}^i \gamma$ will be asymptotically uniformly distributed along the singularities of $\mathcal{F}_Q$. Hence one can express their contribution in terms of some integrals over $\mathcal{S}$. In fact, this idea is realized, in a different situation, in Chapter 5.
- In the analysis of the last term in (4.38), instead of estimating $(Q_m, V_m) - (Q, V)$ by $\mathcal{O}(\varepsilon_\gamma)$ we can use more precise formulas

$$Q_m - Q = \sum_j V_j s_j \quad \text{and} \quad V_m - V \sim \sum_j \frac{(\mathcal{A} \circ \mathcal{F}^j)}{\sqrt{M}},$$

where s_j are intercollision times. Combining these with the Taylor expansion of $A_{Q_m, V_m} - A_{Q, V}$ we obtain a series with many terms of zero mean which lead to additional cancelation. A similar method can be useful to improve our estimates on the terms $I\!I$ and $I\!I\!I$ in (4.42) (this is actually done in [79] in the case of uniformly hyperbolic systems without singularities), but it is technically quite complicated since h and J depend on infinite orbits.

CHAPTER 5

Regularity of the diffusion matrix

5.1. Transport coefficients. In this section we establish the log-Lipschitz continuity, in the sense of (2.16), for the diffusion matrix $\sigma_Q^2(\mathcal{A})$ given by (1.16). Our arguments, however, can be used for the analysis of other transport coefficients in a periodic Lorentz gas (such as electrical conductivity, heat conductivity, viscosity, etc.), so we precede the proof of (2.16) by a general discussion.

Computing transport coefficients is one of the central problems in linear response theory of statistical physics. The evolution of various macroscopic quantities such as mass, momentum, heat, and charge can be described by transport equations, which are very general and can be derived from a few basic principles. They have a wide range of applicability, in the sense that one equation can describe transport in different media. However, the numerical values of transport coefficients are material specific and cannot be found from general principles used to derive the equations themselves. In physics, the values of transport coefficients often have to be determined experimentally. Obtaining the values of transport coefficients theoretically, from the microstructure of the material, seems to be a difficult task.

The difficulty in computing transport coefficients may be partly due to their erratic dependence on the parameters involved. It has been noticed recently that transport coefficients are not differentiable with respect to the model's parameters in several seemingly unrelated cases: one-dimensional piecewise linear mappings [**42, 44, 57, 59, 60**], nonlinear baker transformations [**43**], nonhyperbolic climbing-sine maps [**62**], billiard particles bouncing against a corrugated wall [**45**], and various modifications of a periodic Lorentz gas [**5, 46, 61**]. The only common feature of these models is the presence of singularities in the dynamics. Actually, for completely smooth chaotic systems, such as Anosov diffeomorphisms, the transport coefficients are proven to be differentiable [**78, 79**].

In this section we analyze the diffusion matrix $\sigma_Q^2(\mathcal{A})$ in a periodic Lorentz gas. Even though we only derive an upper bound on its variation, our results and analysis strongly suggest that it may be not differentiable with respect to Q. A similar conjecture was stated in [**5**], where another transport coefficient (electric conductivity) for the periodic Lorentz gas was studied numerically and semi-heuristically. The lack of differentiability of the electric conductivity was traced in [**5**] to singularities in the dynamics, and these are the same singularities that cause divergence of certain terms in our estimates. Eventually we hope to prove rigorously that transport coefficients are not smooth, but so far this remains an open problem. Let us also mention that the regularity of transport coefficients is an issue for stochastic models of interacting particles, see, e.g. [**95**].

Next we describe several specific problems related to transport coefficients. We restrict our discussion to a periodic Lorentz gas with finite horizon; other models are discussed, e.g., in [**6, 47, 90**].

A. DIFFUSION. Consider a single particle moving in a periodic array of scatterers in $\mathbb{R}^2$. Let $q(t)$ denote the position of the particle and $x(t)$ the projection of its position and velocity onto the unit tangent bundle over the fundamental domain (the latter is a torus minus the scatterers). Let x_n be the value of x at the moment of the n-th collision and q_n the position of the particle in $\mathbb{R}^2$ at this moment. Then we have

$$q_n = \sum_{j=0}^{n-1} H(x_j),$$

where $H(x_j)$ denotes the displacement (the change in position) of the particle between the jth and the $(j+1)$st collisions (obviously this difference does not depend on which lift of x to the plane we choose). The Central Limit Theorem for dispersing billiards now gives the following:

THEOREM 4 ([**11**]). *If x_0 has a smooth initial density with respect to the Lebesgue measure, then $q_n/\sqrt{n}$ converges, as $n \to \infty$, to a normal law $\mathcal{N}(0, \bar{D}^2)$ with non-singular covariance matrix $\bar{D}^2$ given by*

$$(5.1) \qquad \bar{D}^2 = \sum_{n=-\infty}^{\infty} \int_{\Omega} H(x_0) H(x_n) \, d\mu(x)$$

where μ denotes the invariant measure on the collision space Ω.

Now standard methods allow us to pass from discrete to continuous time (see [**75, 33, 69**] or our Section 6.7) and we obtain

COROLLARY 5.1 ([**11**]). *If $x(0)$ has a smooth initial density in the phase space, then $q(t)/\sqrt{t}$ converges to $\mathcal{N}(0, D^2)$ where*

$$(5.2) \qquad D^2 = \bar{D}^2/\bar{L}.$$

This result for a single particle system allows us to describe the diffusion in the ideal gas of many noninteracting particles. For example let ρ_0 be a smooth nonnegative function with a compact support. Pick some $\varepsilon > 0$ and for every $m \in \mathbb{Z}^2$ put $N_\varepsilon = [\varepsilon^{-1}\rho_0(\varepsilon m)]$ independent particles into the fundamental domain, which is centered at m, so that each particle's position and velocity direction are uniformly distributed with respect to the Lebesgue measure. Let $\nu_{\varepsilon,t}$ be the measure on $\mathbb{R}^2$ given by $\nu_{\varepsilon,t}(B) = \varepsilon^{-3} \times \#(\text{particles in } B/\varepsilon \text{ at time } t/\varepsilon^2)$. Endow the space of measures with weak topology. Then $\nu_{\varepsilon,t}$ converges in probability, as $\varepsilon \to 0$, to a measure ν_t with density ρ_t, which is the convolution $\rho_t = \rho_0 * \mathcal{N}(0, D^2 t)$, i.e. ρ_t satisfies the diffusion equation

$$\frac{\partial \rho}{\partial t} = \frac{1}{2} \sum_{i,j} D_{ij}^2 \frac{\partial^2 \rho}{\partial y_i \partial y_j},$$

where D^2 is the matrix given by (5.2), and y_1, y_2 denote the coordinates in $\mathbb{R}^2$.

B. ELECTRIC CONDUCTANCE. Consider the previous model with a single particle moving in a periodic array of scatterers, and in addition assume that between collisions the motion is governed by the equation

$$(5.3) \qquad \dot{v} = E - \frac{\langle v, E \rangle}{\|v\|^2} v$$

where $E \in \mathbb{R}^2$ is a fixed vector representing a constant electric field; the second term in (5.3) is the so called Gaussian thermostat, it models the energy dissipation (observe that (5.3) preserves kinetic energy). Let $\mathcal{F}_E : \Omega \to \Omega$ denote the induced collision map.

THEOREM 5 ([23]). (a) *For small E there exists an $\mathcal{F}_E$-invariant ergodic measure μ_E such that for almost all x for all $A \in C(\Omega)$*

$$\frac{1}{n} \sum_{j=0}^{n-1} A(\mathcal{F}_E x) \to \mu_E(A),$$

and μ_E is an SRB measure, i.e. its conditional distributions on unstable manifolds are smooth.
(b) *If A is piecewise Hölder continuous, then*

$$(5.4) \qquad \mu_E(A) = \mu(A) + \omega(A, E) + o(\|E\|)$$

where ω is linear in each variable.

Equation (5.4) is typical for linear response theory in statistical physics – it describes the response of the system to small perturbations of its parameters (here the parameter vector E), up to a linear order.

As before we apply this result to the displacement of the moving particle between consecutive collisions. Part (a) implies that for almost all initial conditions there exists a limit

$$\bar{J}(E) = \lim_{n \to \infty} \frac{q_n}{n} \tag{5.5}$$

which we can interpreted as electrical current (the average speed of the charged particle, see also below). Part (b) implies that there exists a matrix $\bar{M}$ such that for small E

$$\bar{J}(E) = \bar{M}E + o(\|E\|). \tag{5.6}$$

In other words, $\bar{J}$ is a differentiable function of E at $E = 0$. Note, however, that numerical evidence [5] indicates that it is not always differentiable for $E \neq 0$.

As in Subsection A above, (5.5) implies that there exists a limit

$$J(E) = \lim_{t \to \infty} \frac{q(t)}{t} = \frac{\bar{J}(E)}{\bar{L}(E)} \tag{5.7}$$

where $\bar{L}(E)$ denotes the mean (with respect to μ_E) free path. Since $L(E) \to \bar{L}$ as $E \to 0$, it follows that

$$J(E) = \frac{\bar{M}E}{L} + o(\|E\|).$$

Similarly to Subsection A, this result can be applied to an ideal gas. For example consider an infinitely long "wire" W obtained by identifying points of $\mathbb{R}^2$ whose second coordinates differ by an integer. Let $S = \{x = 0\}$ be a vertical line cutting W in half. Put one particle to each fundamental domain in W independently and uniformly distributed with respect to the Lebesgue measure. Let $N_+(t)$ be the number of particles crossing S from left to right during the time interval $(0, t)$, and $N_-(t)$ be the number of particles crossing S from right to left; denote $N(t) = N_+(t) - N_-(t)$. Then (5.7) implies that almost surely there exists a limit

$$\lim_{t \to \infty} \frac{N(t)}{t} = \langle J(E), e_1 \rangle.$$

Thus, the flow of particles in our wire is an electric current which is for small fields approximately proportional to the "voltage" $\langle E, e_1 \rangle$ – we arrive at classical Ohm's law of physics. To compute the coefficient in the corresponding equation, we need to know the functional ω appearing in (5.4). It can be obtained by the following argument (Kawasaki formula):

$$\mu_E(A) = \lim_{n \to \infty} \mu(A \circ \mathcal{F}_E^n),$$

5. REGULARITY OF THE DIFFUSION MATRIX

$$\mu(A \circ \mathcal{F}_E^n) = \mu(A) + \sum_{j=0}^{n-1} \left[\mu(A \circ \mathcal{F}_E^{j+1}) - \mu(A \circ \mathcal{F}_E^j)\right].$$

To estimate the terms in the last sum let $y = \mathcal{F}_E(x)$. Since $\mathcal{F}$ preserves the measure μ, it follows that

$$\frac{d\mu(y)}{d\mu(x)} = 1 + \operatorname{div}_\mu\left(\frac{\partial \mathcal{F}_E}{\partial E}\right) + \mathcal{O}(\|E\|^2).$$

Hence

$$\int A(\mathcal{F}_E^{j+1} x)\, d\mu(x) = \int A(\mathcal{F}_E^j y)\, \frac{d\mu(x)}{d\mu(y)}\, d\mu(y)$$
$$= \int A(\mathcal{F}_E^j y)\left[1 - \operatorname{div}_\mu\left(\frac{\partial \mathcal{F}_E}{\partial E}\right)\right] d\mu(y) + O(\|E\|^2).$$

It follows that

$$\omega(A, E) = -\sum_{j=0}^{\infty} \int \operatorname{div}_\mu\left(\frac{\partial \mathcal{F}_E}{\partial E}\right)(y)\, A(\mathcal{F}^j y)\, d\mu(y)$$

expressing the derivative of μ_E as the sum of correlations. In our case the divergence in question is easy to compute so we obtain the relation $\bar{J} = \frac{1}{2}\bar{D}^2 E$, where $\bar{D}^2$ is the matrix given by (5.1). In other words, $\bar{M} = \frac{1}{2}\bar{D}^2$, and, respectively, $J = \frac{1}{2}D^2 E$, where D^2 is the matrix given by (5.2). This is known in physics as Einstein relation [23].

C. VISCOSITY. This transport coefficient characterizes the flow of momentum in gases. By its very nature, it can only be defined for systems with several interacting particles, so we will not discuss it here (note, however, that a very simplified version of viscosity in a gas with only two molecules was introduced in [13]).

D. RAYLEIGH GAS. So far we have discussed identical particles moving in a periodic configuration of fixed scatterers, but similar considerations apply to Rayleigh gases, in which one or several big massive particles are submerged into an ideal gas of light particles in the open space. Another possibility is to take only one light particle but place it in a semi-open container, such as a halfplane or a section of the plane between two intersecting lines, cf. [14]. In this case an analysis similar to the one given in Section 1.3 leads us to a diffusion equation for the big particle(s), but in contrast with the single particle case [36], a typical light particle collides with several heavy ones before escaping to infinity. So the coefficients of the corresponding transport equations are sum of infinite series. Unfortunately our method cannot be applied to this case yet, because of the lack of necessary results about mixing properties of open billiards, but once such results become available (see [**29**] for a discussion of a simplified model), our method could be used for the study of the well-posedness of transport equations.

We summarize our discussion as follows:

- Transport coefficients are given by infinite correlation sums.
- The regularity of the transport coefficients plays an important role in proving well-posedness of the corresponding transport equations.
- There is an experimental evidence that for billiard problems transport coefficients are not smooth, but this has yet to be established analytically.

We now turn to our primary goal – proving the log-Lipschitz continuity of the diffusion matrix, as claimed by (2.16). Let A and B be smooth functions on the $r\varphi$ coordinate chart Ω_0 such that $\int A\, d\mu_0 = \int B\, d\mu_0 = 0$ (in fact, it is enough that one integral vanishes). Recall that the spaces Ω_Q are identified with the Ω_0, hence our functions A, B are defined on Ω_Q. For every $Q \in \mathcal{D}$ such that $\text{dist}(Q, \partial\mathcal{D}) \geq \mathbf{r} + \delta$ we put

$$(5.8) \qquad \bar{\sigma}_Q^2(A, B) = \sum_{j=-\infty}^{\infty} \int_{\Omega_Q} A \left(B \circ \mathcal{F}_Q^j \right) d\mu_Q.$$

(Here we let j change from $-\infty$ to $+\infty$ but of course our result is valid for one sided sums as well).

PROPOSITION 5.2. *Under Assumption A3', for all $Q_1 \approx Q_2$ we have*

$$(5.9) \qquad \left| \bar{\sigma}_{Q_1}^2(A, B) - \bar{\sigma}_{Q_2}^2(A, B) \right| \leq \text{Const} \, \|Q_1 - Q_2\| \left| \ln \|Q_1 - Q_2\| \right|$$

The bound (5.9), along with Assumption A4 on the nonsingularity of the matrix $\sigma_Q^2(\mathcal{A})$, immediately implies the required (2.16), so it remains to prove Proposition 5.2.

5.2. Reduction to a finite series. We first discuss our approach to the problem. For smooth uniformly hyperbolic systems, the dynamical invariants, such as diffusion coefficients, are usually differentiable with respect to the parameters of the model [**52**]. In dispersing billiards, the presence of singularities makes the dynamics *nonuniformly* hyperbolic, and for such systems, no similar results are available. On the contrary, there is an experimental evidence (supported by heuristic arguments) that dynamical invariants are, generally, not differentiable, see [**5**]. Our proposition is the first positive result in this direction.

5. REGULARITY OF THE DIFFUSION MATRIX

Due to the identification of Ω_Q with Ω_0, all the maps $\mathcal{F}_Q$ act on the same space Ω_0 and preserve the same measure $d\mu_0 = c^{-1}\cos\varphi\, dr\, d\varphi$, where $c = 2\,\text{length}(\partial\mathcal{D}) + 4\pi\mathbf{r}$ is the normalizing factor. Hence we can treat $\mathcal{F}_{Q_2}$ as a perturbation of $\mathcal{F}_{Q_1}$ in (2.16).

There are two equivalent approaches to establish the regularity of dynamical invariants for hyperbolic maps under perturbations. The analytic approach consists of term-by-term differentiation of the relevant infinite series, like our (5.8), with respect to the parameters of the model (in our case, it is Q) and then integrating by parts. The geometric approach is based on an explicit comparison of the orbits under the two maps and using a shadowing-type argument.

The analytic method is shorter, if somewhat less transparent, since it involves algebraic manipulations instead of geometric considerations. However its range of applicability is rather narrow, because it requires differentiability at all the relevant values of parameters, whereas the geometric method is more flexible and may handle less regular parameterizations. For this reason we have used the geometric approach in the proof of Proposition 3.3 because we treated the dynamics as a small perturbation of the $M = \infty$ case, where we had no analyticity (in M). For the proof of Proposition 5.2 we choose the analytic method, but we hope that after the proof of Proposition 3.3 the geometric meaning of our manipulations is clear.

The main difference between the proofs of Propositions 5.2 and 3.3 is that in the latter we had a luxury of discarding orbits which come close to the singularities, but now we have to take them into account as well. The contribution of those orbits is described by certain triple correlation functions

$$\beta_{m,n} = \int_S (A \circ \mathcal{F}_Q^{-m})\, B\, (C \circ \mathcal{F}_Q^n)\, d\nu$$

where A, B, C are smooth functions and the measure ν is concentrated on a singularity curve S of the map $\mathcal{F}_Q$. If ν were a smooth measure on the entire space Ω_0, then we could use a local product structure to show, in a usual way, an asymptotic independence of the future and the past, and the triple correlations could be bounded as in [21], so that $\sum_{m,n\geq 0} |\beta_{m,n}| < \infty$. However, ν is concentrated on a curve S, which has no local product structure, and the series $\sum_{m,n\geq 0} |\beta_{m,n}|$ appears to diverge. In fact, we are only able to get an estimate growing with the number of terms: $\sum_{m+n\leq N} |\beta_{m,n}| = \mathcal{O}(N)$, and it is this estimate that gives us the logarithmic factor in (5.9).

For brevity, we will write $\Omega = \Omega_0$ and $\mu = \mu_0$. Let

$$D_{A,B}^{(N)}(Q) = \sum_{n=1}^{N} \mu\left[A\left(B \circ \mathcal{F}_Q^n\right)\right]$$

Our main estimate is

(5.10) $$\left|\frac{dD_{A,B}^{(N)}(Q)}{dQ}\right| \leq \mathrm{Const}_{A,B}\, N$$

where d/dQ denotes the directional derivative along an arbitrary unit vector in the Q plane.

Proof of Proposition 5.2. According to a uniform exponential bound on correlations (Extension 1 in Section A.1),

$$\left|\mu\left[A\left(B \circ \mathcal{F}_Q^n\right)\right]\right| \leq \mathrm{Const}_{A,B}\, \theta^{|n|}$$

for some $\theta < 1$. Let $N = 2\ln\|Q_1 - Q_2\|/\ln\theta$. Then for any Q

$$\bar\sigma_Q^2 = \mu(AB) + D_{A,B}^{(N)}(Q) + D_{B,A}^{(N)}(Q) + \mathcal{O}\left(\|Q_1 - Q_2\|^2\right)$$

where the first term does not depend on Q. Hence

$$\bar\sigma_{Q_1}^2 - \bar\sigma_{Q_2}^2 = \left[D_{A,B}^{(N)}(Q_1) + D_{B,A}^{(N)}(Q_1) - D_{A,B}^{(N)}(Q_2) - D_{B,A}^{(N)}(Q_2)\right]$$
$$+ \mathcal{O}\left(\|Q_1 - Q_2\|^2\right)$$

The main estimate (5.10) implies that the expression in brackets is bounded by $\mathrm{Const}_{A,B}\|Q_1 - Q_2\|\, N$. □

The rest of this section is devoted to proving the main estimate (5.10).

5.3. Integral estimates: general scheme. Let

$$I_n = \frac{d}{dQ}\mu\left[A\left(B \circ \mathcal{F}_Q^n\right)\right]$$

Since the rest of the proof deals with $\mathcal{F}_Q$ for a fixed Q we shall omit the subscript from now on. We also put $A_n = A \circ \mathcal{F}^n$. Note that A is smooth on Ω but A_n has discontinuities on the singularity set $\mathcal{S}_n \subset \Omega$ of the map $\mathcal{F}^n$. The curves of $\mathcal{S}_n$ change with Q smoothly, so we have

$$I_n = I_n^{(c)} + I_n^{(d)}$$

where the first term contains the derivative of the integrand

$$I_n^{(c)} = \int_\Omega A \frac{dB_n}{dQ}\, d\mu$$

5. REGULARITY OF THE DIFFUSION MATRIX

and the second one contains the boundary integrals

$$(5.11) \qquad I_n^{(d)} = \int_{\mathcal{S}_n \setminus \mathcal{S}_0} A\left(\Delta B_n\right) v^\perp \cos\varphi \, dl$$

where ΔB_n denotes the jump of B_n across $\mathcal{S}_n \setminus \mathcal{S}_0$, $v^\perp$ is the velocity of $\mathcal{S}_n \setminus \mathcal{S}_0$ as it changes with Q (in the normal direction), and dl the Lebesgue measure (length) on $\mathcal{S}_n$. Observe that $\mathcal{S}_0 = \partial\Omega$ does not change with Q, hence it need not be included in $I_n^{(d)}$. We postpone the analysis of the boundary terms until Section 5.5.

Now consider a vector field on Ω defined by

$$X = \frac{d\mathcal{F}}{dQ} \circ \mathcal{F}^{-1}$$

For an $x = (r, \varphi)$ such that either x or $\mathcal{F}^{-1}(x)$ lies on $\partial\mathcal{P}(Q) \times [-\pi/2, \pi/2]$ X vanishes, otherwise X is an unstable vector with coordinates

$$(5.12) \qquad X = (dr_X, d\varphi_X) = \left(\frac{\sin(\varphi+\psi)}{\cos\varphi}, \mathcal{K}\frac{\sin(\varphi+\psi)}{\cos\varphi}\right)$$

where $\mathcal{K} > 0$ is the curvature of the boundary $\partial\mathcal{D} \cup \partial\mathcal{P}(Q)$ at x, and ψ is the angle between the normal to the boundary at the point x and the direction of our derivative d/dQ. Note that $\|X\| = \mathcal{O}(1/\cos\varphi)$ is unbounded but $\mu(\|X\|) < \infty$, because the density of μ is proportional to $\cos\varphi$. It is also easy to check that $\mu(\|d\mathcal{F}^k(X)\|) < \infty$ for all $k \geq 1$.

It is now immediate that

$$I_n^{(c)} = \sum_{k=0}^{n-1} I_{n,k}^{(c)}$$

where

$$(5.13) \qquad I_{n,k}^{(c)} = \int_\Omega A\left[\left(\partial_{d\mathcal{F}^k(X)} B\right) \circ \mathcal{F}^n\right] d\mu$$

Note that $d\mathcal{F}^k(X)$ grows exponentially fast with k. To properly handle these integrals, we will decompose $d\mathcal{F}^k(X)$ into stable and unstable components.

Let $E^u = E^u(x)$ be the field of unstable directions given by equation $d\varphi/dr = \mathcal{K}$, and $E^s = E^s(x)$ be the field of stable directions given by equation $d\varphi/dr = -\mathcal{K}$. The field E^u corresponds to (infinitesimal) families of trajectories that are parallel before the collision at x, and E^s corresponds to families of trajectories that are parallel after the collision. Note that in contrast with more common notation E^u and E^s are not invariant under dynamics. Rather they are smooth vector fields such that $d\mathcal{F}(E^s) = E^u$ and X belongs in E^u.

Let $\mathcal{G}$ be a smooth foliation of Ω by u-curves that integrate the field E^u. Then $\mathcal{G}_m = \mathcal{F}^m(\mathcal{G})$, for $m \geq 0$, is a piecewise smooth foliation by u-curves that integrate the field $E^u_m = d\mathcal{F}^m(E^u)$. Note that the discontinuities of $\mathcal{G}_m$ coincide with those of the map $\mathcal{F}^{-m}$. Let Π^u_m and Π^s_m denote the projections onto E^u_m and E^s, respectively, along E^s and E^u_m. Let $\Theta^*_m = \Pi^*_m \circ d\mathcal{F}$, where $* = u, s$. For $k > m \geq 0$, let

$$\Theta^s_{m,k} = \Theta^s_k \circ \cdots \circ \Theta^s_{m+2} \circ \Theta^s_{m+1}$$

LEMMA 5.3. *There is a constant $\theta < 1$ and a function $u(x)$ on Ω such that for any nonzero vector $dx \in E^s$ and $m \geq 0$ we have*

(5.14) $$\frac{\|\Theta^s_m(dx)\|}{\|dx\|} \leq \theta \frac{u(\mathcal{F}(x))}{u(x)}$$

The function $u(x)$ is bounded away from zero and infinity:

$$0 < u_{\min} < u(x) < u_{\max} < \infty,$$

therefore, for any $0 \neq dx \in E^s$

(5.15) $$\frac{\|\Theta^s_{m,k}(dx)\|}{\|dx\|} \leq \theta^{k-m} \frac{u(\mathcal{F}^{k-m}(x))}{u(x)} \leq \theta^{k-m} \frac{u_{\max}}{u_{\min}}$$

for any $k > m$. Lastly, $\|\Pi^s_m(X)\| \leq$ Const.

Proof. We denote $x = (r, \varphi)$ and $\mathcal{F}(x) = x_1 = (r_1, \varphi_1)$. Note that the vectors $dx = (dr, d\varphi) \in E^s(x)$ and $d\mathcal{F}(dx) = dx_1 = (dr_1, d\varphi_1) \in E^u(x_1)$ correspond to an (infinitesimal) family of trajectories that remain parallel between the collisions at x and x_1, and this family is characterized by the vector dq (orthogonal to the velocity vector) and $dv = 0$, in the notation of Section 4.1, and we have

$$\|dq\| = |\cos\varphi\, dr| = |\cos\varphi_1\, dr_1|$$

Recall that the norm of s-vectors is defined by (4.15), where $\mathbf{s}_V = 1$ since $V = 0$, hence

$$\|dx\|^2 = (4\mathcal{K}^2 + \cos^2\varphi)(dr)^2$$

and for the s-vector $\Theta^s_m(dx) = dx^s_1 = (dr^s_1, d\varphi^s_1) \in E^s(x_1)$ we have

$$\|dx^s_1\|^2 = (4\mathcal{K}^2_1 + \cos^2\varphi_1)(dr^s_1)^2$$

where $\mathcal{K}$ and $\mathcal{K}_1$ denote the curvature of the boundary at x and x_1, respectively. Next, the vector dx^s_1 is the projection of dx_1 onto $E^s(x_1)$ along $E^u_m(x_1)$, the latter is given by equation $d\varphi/dr = \mathcal{K}_1 + \mathcal{B}_1 \cos\varphi_1$, where $\mathcal{B}_1$ is the curvature of the precollisional family of trajectories corresponding to $E^u_m(x_1)$. By a direct inspection, see Fig. 1, we have

(5.16) $$|dr^s_1| = |dr_1| \frac{\mathcal{B}_1 \cos\varphi_1}{2\mathcal{K}_1 + \mathcal{B}_1 \cos\varphi_1}$$

5. REGULARITY OF THE DIFFUSION MATRIX

hence

$$\text{(5.17)} \quad \frac{\|dx_1^s\|^2}{\|dx\|^2} = \frac{4\mathcal{K}_1^2 + \cos^2 \varphi_1}{4\mathcal{K}^2 + \cos^2 \varphi} \times \frac{\cos^2 \varphi}{(2\mathcal{B}_1^{-1}\mathcal{K}_1 + \cos \varphi_1)^2}$$

Note that $0 < \mathcal{B}_1 \leq 1/s$, where s the free path length between the points x and x_1, hence $\mathcal{B}_1 \leq 1/L_{\min}$. Thus

$$\text{(5.18)} \quad \frac{\|dx_1^s\|^2}{\|dx\|^2} \leq \frac{4\mathcal{K}_1^2 + \cos^2 \varphi_1}{4\mathcal{K}^2 + \cos^2 \varphi} \times \frac{(c_0 + \cos \varphi)^2}{(c_0 + \cos \varphi_1)^2} \times \frac{\cos^2 \varphi}{(c_0 + \cos \varphi)^2}$$

where $c_0 = 2L_{\min}\mathcal{K}_{\min} > 0$. Now we put $u(x) = (4\mathcal{K}^2 + \cos^2 \varphi)^{1/2}/(c_0 + \cos \varphi)$ and $\theta = 1/(c_0 + 1)$, which proves (5.14). Replacing x_1 by x and dx_1 by $X = (dr_X, d\varphi_X)$, see (5.12), in the above argument gives an estimate for the vector $\Pi_m^s(X) = (dr^s, d\varphi^s)$:

$$\|\Pi_m^s(X)\|^2 = (4\mathcal{K}^2 + \cos^2 \varphi)(dr^s)^2$$
$$= \frac{(\mathcal{B} \cos \varphi)^2}{(2\mathcal{K} + \mathcal{B} \cos \varphi)^2}(4\mathcal{K}^2 + \cos^2 \varphi)(dr_X)^2$$
$$\leq \frac{4\mathcal{K}_{\max}^2 + 1}{4\mathcal{K}_{\min}^2/\mathcal{B}_{\max}^2} \qquad \square$$

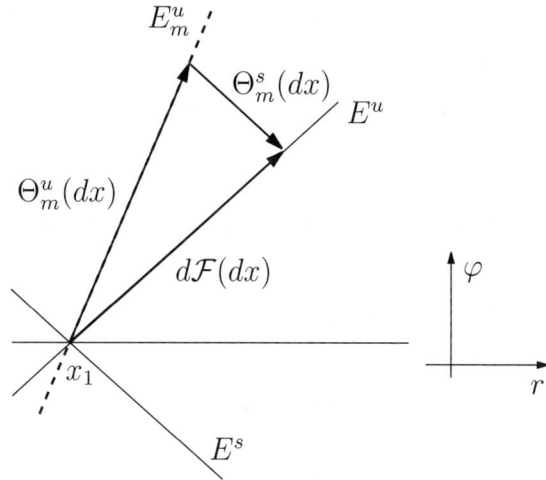

FIGURE 1. The decomposition $d\mathcal{F}(dx) = \Theta_m^u(dx) + \Theta_m^s(dx)$

Remark. As (5.16) implies,

$$\|d\mathcal{F}(dx)\| \leq \text{Const} \frac{\|\Theta_m^s(dx)\|}{\cos \varphi_1}$$

hence

$$(5.19) \quad \frac{\|\Theta^u_{k+1} \circ \Theta^s_{m,k}(dx)\|}{\|dx\|} \leq \text{Const} \frac{\theta^{k-m}}{\cos \varphi_{k-m+1}}$$

where we denote $\mathcal{F}^{k-m+1}(x) = (r_{k-m+1}, \varphi_{k-m+1})$. Since $\mathcal{F}^{-1}$ uniformly contracts u-vectors by a factor $\mathcal{O}(\cos \varphi)$, then

$$(5.20) \quad \frac{\|d\mathcal{F}^{-1} \circ \Theta^u_{k+1} \circ \Theta^s_{m,k}(dx)\|}{\|dx\|} \leq \text{Const } \theta^{k-m}$$

Remark. For a future reference, we record a slight improvement of the estimate (5.14):

$$(5.21) \quad \forall dx \in E^s \quad \frac{\|\Theta^s_m(dx)\|}{\|dx\|} \leq \theta \, R(\cos \varphi) \frac{u(\mathcal{F}(x))}{u(x)}$$

where $R(\cos \varphi) = \min\{1, C_0 \cos \varphi\}$ and $C_0 = 1 + c_0^{-1}$. This improvement follows from (5.18).

We now return to the integral (5.13). Let us decompose

$$X = \bar{\alpha}_{n-k,0} + \beta_{n-k,0}$$

where

$$\bar{\alpha}_{n-k,0} = \Pi^u_{n-k}(X) \in E^u_{n-k}, \qquad \beta_{n-k,0} = \Pi^s_{n-k}(X) \in E^s$$

and, inductively, for $r \geq 1$,

$$d\mathcal{F}^r(X) = \bar{\alpha}_{n-k,r} + \beta_{n-k,r}$$

where

$$\bar{\alpha}_{n-k,r} = d\mathcal{F}(\bar{\alpha}_{n-k,r-1}) + \Theta^u_{n-k+r}(\beta_{n-k,r-1}) \in E^u_{n-k+r}$$

and

$$\beta_{n-k,r} = \Theta^s_{n-k+r}(\beta_{n-k,r-1}) \in E^s$$

Observe that

$$\beta_{n-k,k} = \Theta^s_{n-k,n}(\beta_{n-k,0})$$

and

$$\bar{\alpha}_{n-k,k} = d\mathcal{F}^k(\bar{\alpha}_{n-k,0}) + \sum_{j=0}^{k-1} d\mathcal{F}^j \circ \Theta^u_{n-j}(\beta_{n-k,k-j-1})$$

Denote

$$\alpha_{n-k,k-j} = \Theta^u_{n-j}(\beta_{n-k,k-j-1}) \in E^u_{n-j}$$

If we also put, for convenience of notation, $\alpha_{n-k,0} = \bar{\alpha}_{n-k,0}$ and denote

$$\alpha^{(m)}_{n-k,k-j} = d\mathcal{F}^m(\alpha_{n-k,k-j}) \quad \forall \, m \in \mathbb{Z}$$

5. REGULARITY OF THE DIFFUSION MATRIX

then we obtain

(5.22) $$d\mathcal{F}^k(X) = \sum_{j=0}^{k} \alpha_{n-k,k-j}^{(j)} + \beta_{n-k,k}$$

Accordingly,

$$I_{n,k}^{(c)} = \sum_{j=0}^{k} I_{n,k,j}^{(u)} + I_{n,k}^{(s)}$$

where

(5.23) $$I_{n,k,j}^{(u)} = \int_\Omega A\left[(\partial_{\alpha_{n-k,k-j}^{(j)}} B) \circ \mathcal{F}^n\right] d\mu$$

and

(5.24) $$I_{n,k}^{(s)} = \int_\Omega A\left[(\partial_{\beta_{n-k,k}} B) \circ \mathcal{F}^n\right] d\mu$$

LEMMA 5.4. *There is a constant $\theta < 1$ such that for all $0 \le k < n$*

$$\|\alpha_{n-k,k-j}\| \le \text{Const}\, \theta^{k-j}/\cos\varphi,$$
$$\|\alpha_{n-k,k-j}^{(-m)}\| \le \text{Const}\, \theta^{k-j+m} \quad \forall\, m \ge 1$$

and

$$\|\beta_{n-k,k}\| \le \text{Const}\, \theta^k.$$

Proof. Use Lemma 5.3 and the subsequent remark. □

COROLLARY 5.5.
$$\left|\sum_{n=1}^{N}\sum_{k=0}^{n-1} I_{n,k}^{(s)}\right| \le \text{Const}\, N.$$

Proof. Estimating the integrand in (5.24) by its absolute value we get $\left|I_{n,k}^{(s)}\right| \le \text{Const}\, \|A\|_\infty \|B\|_{C^1} \theta^k.$ □

5.4. Integration by parts. The estimation of $I_{n,k,j}^{(u)}$ in (5.23) requires integration by parts. Changing variables $y = \mathcal{F}^{n-j-1}x$ gives

(5.25) $$I_{n,k,j}^{(u)} = \int_\Omega A_{-(n-j-1)}\left(\partial_{\alpha_{n-k,k-j}^{(-1)}} B_{j+1}\right) d\mu$$

(we have to work with $d\mathcal{F}^{-1}\alpha_{n-k,k-j}$, instead of $\alpha_{n-k,k-j}$ to avoid an infinite growth of the latter as $\cos\varphi \to 0$, see Lemma 5.4).

Observe that the integrand in (5.25) is discontinuous on the set $\mathcal{S}_{-(n-j-1)}$ (due to $A_{-(n-j-1)}$ and the vector field) and $\mathcal{S}_{j+1}$ (due to B_{j+1}), hence we have to integrate by parts on each connected domain $D \subset \Omega \setminus (\mathcal{S}_{-(n-j-1)} \cup \mathcal{S}_{j+1})$.

Observe that $\alpha_{n-k,k-j}^{(-1)} \in E_{n-j-1}^u$, hence the integral curves of this vector field are the fibers of the foliation $\mathcal{G}_{n-j-1}$. For every domain D, denote by $\mathcal{G}_D = \{\gamma_D\}$ the fibers of this foliation restricted to D and by ρ the densities of the corresponding conditional measures on them. Let λ_D denote the factor measure. To simplify our notation, we put $A_- = A_{-(n-j-1)}$, $B_+ = B_{j+1}$ and $\alpha = \|\alpha_{n-k,k-j}^{(-1)}\|$. For any curve γ, let $\int_\gamma C\,dx$ denote the integral of a function C with respect to the arclength parameter on γ, and $C' = \partial C/\partial x$ denote the derivative along γ. Then the integration by parts gives

$$I_{n,k,j}^{(u)} = \sum_D \int_{\mathcal{G}_D} d\lambda_D \int_{\gamma_D} A_-\,\alpha\,B_+'\,\rho\,dx$$

(5.26)
$$= I_{n,k,j}^{(b)} - I_{n,k,j}^{(i)}$$

where

(5.27) $$I_{n,k,j}^{(b)} = \int_{(\mathcal{S}_{-(n-j-1)} \cup \mathcal{S}_{j+1})\setminus \mathcal{S}_0} \Delta\left[A_-\,B_+\,\|\alpha^\perp\|_0\right]\cos\varphi\,dl$$

is the boundary term, which will be analyzed in the next subsection (note that we exclude the set $\mathcal{S}_0 = \{(r,\varphi): \cos\varphi = 0\}$ since $\rho = 0$ on $\mathcal{S}_0$) and

(5.28) $$I_{n,k,j}^{(i)} = \sum_D \int_{\mathcal{G}_D} d\lambda_D \int_{\gamma_D} (\rho\,\alpha\,A_-)'\,B_+\,dx.$$

Observe that

$$(\rho\,\alpha\,A_-)' = A_-'\,\alpha\,\rho + A_-\,\alpha\,(\ln\rho)'\,\rho + A_-\,\alpha'\,\rho,$$

hence the last sum in (5.26) equals

$$\int_\Omega \alpha\,A_-'\,B_+\,d\mu + \int_\Omega \alpha\,(\ln\rho)'\,A_-\,B_+\,d\mu + \int_\Omega \alpha'\,A_-\,B_+\,d\mu.$$

These integrals will be estimated in the next two lemmas.

LEMMA 5.6. *For some constant $\theta < 1$, we have*

(5.29) $$\left|\int_\Omega \alpha\,A_-'\,B_+\,d\mu\right| \leq \text{Const }\theta^{n-j},$$

(5.30) $$\left|\int_\Omega \alpha\,(\ln\rho)'\,A_-\,B_+\,d\mu\right| \leq \text{Const }\theta^{k-j},$$

(5.31) $$\left|\int_\Omega \alpha'\,A_-\,B_+\,d\mu\right| \leq \text{Const }\theta^{k-j}.$$

5. REGULARITY OF THE DIFFUSION MATRIX

Proof. Since $\mathcal{F}^{-1}$ contracts unstable curves by a factor $\leq \vartheta < 1$, we have $\|A'_-\| \leq \text{Const}\, \vartheta^{n-j}$, which proves (5.29).

Lemma 5.4 implies that $\alpha \leq \text{Const}\, \theta^{k-j}$, and so
$$\left| \int_\Omega \alpha\, (\ln \rho)'\, A_-\, B_+\, d\mu \right| \leq \text{Const}\, \theta^{k-j} \int_\Omega |(\ln \rho)'|\, d\mu$$
To show that the last integral is finite, we first need to refine our foliations $\mathcal{G}_m$, $m \geq 0$. We divide the fibers of the original foliation $\mathcal{G}$ into H-curves (by cutting them at the boundaries of the homogeneity strips) and denote the resulting family of H-curves by $\tilde{\mathcal{G}}$. For $m \geq 0$, let $\tilde{\mathcal{G}}_m$ denote the foliation of Ω into the H-components of the sets $\mathcal{F}^m(\gamma)$, $\gamma \in \tilde{\mathcal{G}}$, see Section 4.3 (note that $\tilde{\mathcal{G}}_m$ is a refinement of $\mathcal{G}_m$). Denote by $\gamma_m(x)$ the fiber of $\tilde{\mathcal{G}}_m$ that contains the point x. Now by (B.7)

$$(5.32) \qquad \left| [\ln \rho(x)]' \right| \leq \frac{\text{Const}}{|\gamma_{n-j-1}(x)|^{2/3}}$$

Lemma 4.10 implies that $\mu\{x\colon |\gamma_m(x)| < t\} \leq \text{Const}\, t$ for every $m \geq 0$, hence
$$\int_\Omega \frac{1}{|\gamma_{n-j-1}(x)|^{2/3}}\, d\mu \leq \text{Const} \int_0^1 t^{-2/3}\, dt \leq \text{Const}$$
which proves (5.30). To derive (5.31) we will show that

$$(5.33) \qquad |\alpha'(x)| \leq \frac{\text{Const}\, \theta^{k-j}}{|\gamma_{n-j}(\mathcal{F}(x))|^{2/3}}$$

where $\theta < 1$ is a constant. Then (5.31) will follow by
$$\int_\Omega \frac{\theta^{k-j}}{|\gamma_{n-j}(\mathcal{F}(x))|^{2/3}}\, d\mu \leq \text{Const}\, \theta^{k-j} \int_0^1 t^{-2/3}\, dt \leq \text{Const}\, \theta^{k-j}$$
where we changed variable $y = \mathcal{F}(x)$ and used the invariance of μ.

It remains to prove (5.33). For $j = k$, we have $\alpha = \|\alpha_{n-k,0}^{(-1)}\|$. The vector $\alpha_{n-k,0}^{(-1)}$ is the projection of $d\mathcal{F}^{-1}X$ onto E^u_{n-k-1} along $E^s_{-1}\colon = d\mathcal{F}^{-1}(E^s)$. Similarly to (5.12), we have
$$d\mathcal{F}^{-1}X = (dr, d\varphi) = \left(\frac{\sin(\varphi - \psi)}{\cos \varphi},\, -\mathcal{K}\, \frac{\sin(\varphi - \psi)}{\cos \varphi} \right)$$
hence the vector field $(\cos \varphi)\, d\mathcal{F}^{-1}X$ is C^2 smooth, with uniformly bounded first and second derivatives on Ω. For brevity, we will say that a function is *uniformly C^2 smooth*, if its first and second derivatives are bounded by some constants determined by the domain $\mathcal{D}$, by δ_0 in (3.3), and by our functions A and B. The field E^u_{n-k-1} is uniformly C^2 smooth along the fibers of $\tilde{\mathcal{G}}_{n-k-1}$ by Proposition B.1. The field E^s is given by equation $d\varphi/dr = -\mathcal{K}$, so it is uniformly C^2 smooth on Ω. By

using basic facts about billiards, cf. Appendices A and B, and direct calculation we find that the vector field E^s_{-1} is given by equation

$$d\varphi/dr = -\mathcal{K} - 2\mathcal{K}_1 \cos\varphi/(2s\mathcal{K}_1 + \cos\varphi_1)$$

where $x_1 = (r_1, \varphi_1) = \mathcal{F}(x)$ and $\mathcal{K}_1$ denotes the curvature of the boundary at the point x_1. Note that the lines $E^s(x)$ and $E^s_{-1}(x)$ have slopes bounded away from 0 and $-\infty$, and the difference between these slopes is

(5.34) $$\measuredangle\big(E^s(x), E^s_{-1}(x)\big) = \mathcal{O}(\cos\varphi)$$

If $\hat{\gamma}(x)$ denotes the angle between $E^s(x)$ and $E^s_{-1}(x)$, then $(\cos\varphi)^{-1}\hat{\gamma}(x)$ can be given by a formal expression (in terms of x and x_1) that would be a uniformly C^2 smooth function of x and x_1. However, if we differentiate $(\cos\varphi)^{-1}\hat{\gamma}(x)$ with respect to x along the fibers of the unstable foliation $\tilde{\mathcal{G}}_{n-k-1}$, then x_1 becomes a function of x such that $|dx_1/dx| = \mathcal{J}(x) = \mathcal{O}(1/\cos\varphi_1)$, where $\mathcal{J}(x)$ is the Jacobian of the map $\mathcal{F}\colon \gamma_{n-k-1}(x) \to \gamma_{n-k}(x_1)$. Hence

$$\left|\frac{d}{dx}\big[(\cos\varphi)^{-1}\hat{\gamma}(x)\big]\right| \leq \frac{\text{Const}}{\cos\varphi_1}$$

This gives us an estimate for the derivative along the fibers of $\tilde{\mathcal{G}}_{n-k-1}$:

(5.35) $$\left|\frac{d\alpha(x)}{dx}\right| = \left|\frac{d\alpha(x)}{dx_1}\frac{dx_1}{dx}\right| \leq \frac{\text{Const}}{\cos\varphi_1} \leq \frac{\text{Const}}{|\gamma_{n-k}(\mathcal{F}(x))|^{2/3}}$$

where the last inequality follows from (4.17), which proves (5.33) for $j = k$. For a future reference, we also note that

(5.36) $$\left|\frac{d^2\alpha(x)}{dx^2}\right| = \left|\frac{d[d\alpha/dx]}{dx_1}\frac{dx_1}{dx}\right| \leq \frac{\text{Const}\,\mathcal{J}(x)}{|\gamma_{n-k}(\mathcal{F}(x))|^{4/3}}$$

To prove (5.33) for $j < k$ we use induction on $t := k - j$. Let $\alpha_t = \|\alpha^{(-1)}_{n-k,t}\|$ and $\beta_t = \|\beta_{n-k,t}\|$. Consider the trajectory $x_t = (r_t, \varphi_t) = \mathcal{F}^t(x)$ of a point x. Observe that the vector $\alpha^{(-1)}_{n-k,t}$ is parallel to the line $E^u_{n-k+t-1}$; and α_t, β_{t-1} are two sides of a triangle (shaded on Fig. 2), in which one angle is $\mathcal{O}(\cos\varphi_{t-1})$, cf. (5.34). Therefore,

(5.37) $$\alpha_t = E_{t-1}\beta_{t-1}\cos\varphi_{t-1}$$

where the factor E_{t-1} is bounded away from zero and infinity:

$$0 < E_{\min} \leq E_{t-1} \leq E_{\max} < \infty;$$

and E_{t-1} can be given by a formal expression (in terms of x_{t-1}, x_t, and the slope $\Gamma^u_{t-1} = d\varphi/dr$ of the line $E^u_{n-k+t-1}$ at the point x_{t-1}) that

5. REGULARITY OF THE DIFFUSION MATRIX 73

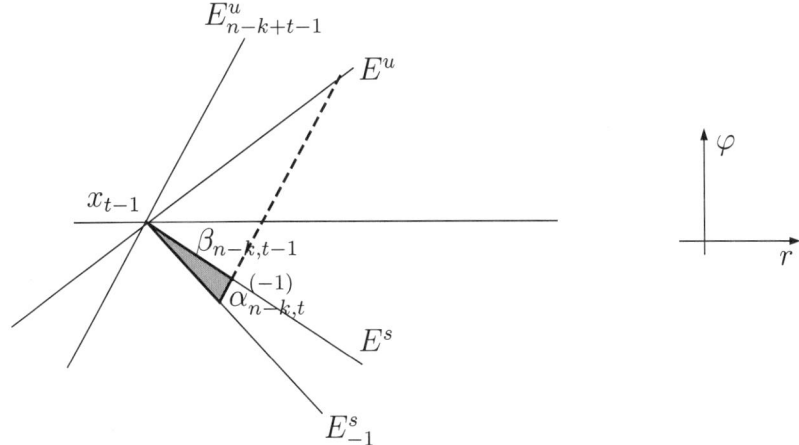

FIGURE 2. The vector $\alpha_{n-k,t}^{(-1)}$ is parallel to $E^u_{n-k+t-1}$

would be a uniformly C^2 smooth function of the variables x_{t-1}, x_t, and Γ^u_{t-1}. Now we have

(5.38) $$\alpha_{t+1} = G_t \alpha_t, \qquad G_t := \frac{E_t \beta_t \cos \varphi_t}{E_{t-1} \beta_{t-1} \cos \varphi_{t-1}}.$$

It follows from (5.17) that

$$H_t := \frac{\beta_t}{\beta_{t-1} \cos \varphi_{t-1}}$$

is bounded away from zero and infinity and can be given by a formal expression (in terms of x_{t-1}, x_t, and the curvature $\mathcal{B}_t$ of the precollisional family of trajectories corresponding to E^u_{n-k+t} at the point x_t) that would be a uniformly C^2 smooth function of x_{t-1}, x_t, and $\mathcal{B}_t$. Then

(5.39) $$G_t = \frac{E_t H_t \cos \varphi_t}{E_{t-1}}$$

is a uniformly C^2 smooth function of x_{t-1}, x_t, x_{t+1}, Γ_{t-1}, Γ_t, and $\mathcal{B}_t$ (the variable x_{t+1} comes only from E_t). We note that Γ_{t-1} and Γ_t are C^2 smooth functions of x_{t-1} and x_t, respectively, and $\mathcal{B}_t$ is a uniformly C^2 smooth function along the corresponding fiber of the foliation $\mathcal{G}_{n-k+t}$, see Appendix B.

Now we differentiate (5.39) with respect to x_t along the corresponding fiber of the foliation $\mathcal{G}_{n-k+t}$ and use $|dx_{t+1}/dx_t| = \mathcal{O}(1/\cos \varphi_{t+1})$

and $|dx_{t-1}/dx_t| < 1$ to obtain

$$\left|\frac{dG_t}{dx_t}\right| \leq \frac{\bar{C}}{\cos\varphi_{t+1}} \tag{5.40}$$

with a constant $\bar{C} > 0$. Next, differentiating the identity $\alpha_{t+1} = G_t \alpha_t$ gives

$$\left|\frac{d\alpha_{t+1}}{dx_t}\right| \leq \frac{\bar{C}\alpha_t}{\cos\varphi_{t+1}} + \left|\frac{E_t \beta_t \cos\varphi_t}{E_{t-1}\beta_{t-1}\cos\varphi_{t-1}} \frac{1}{\mathcal{J}_{t-1}} \frac{d\alpha_t}{dx_{t-1}}\right| \tag{5.41}$$

where $\mathcal{J}_{t-1} = dx_t/dx_{t-1}$ is the Jacobian of the map $\mathcal{F}: \gamma_{n-k+t-1}(x_{t-1}) \to \gamma_{n-k+t}(x_t)$. Note that by (5.21),

$$\frac{\beta_t}{\beta_{t-1}} \leq \frac{u_t}{u_{t-1}} R(\cos\varphi_{t-1}) \tag{5.42}$$

where $u_t = u(x_t)$. Now we prove, by induction on t that there exists a constant $\theta < 1$ and large constants $P, Q > 0$ such that

$$\left|\frac{d\alpha_t}{dx_{t-1}}\right| \leq \frac{E_{t-1} u_{t-1} \cos\varphi_{t-1}}{R(\cos\varphi_{t-1})} \left(P + \frac{Q}{\cos\varphi_t}\right) \theta^t \tag{5.43}$$

Note that the first factor here is bounded away from zero and infinity:

$$0 < \frac{E_{\min} u_{\min}}{C_0} \leq \frac{E_{t-1} u_{t-1} \cos\varphi_{t-1}}{R(\cos\varphi_{t-1})} \leq E_{\max} u_{\max} < \infty$$

For $t = 0$, the bound (5.43) follows from (5.35). (Note that the same angle is denoted by φ_1 in (5.35) and by $\varphi_t = \varphi_0$ in (5.43) with $t = 0$.) Combining (5.41)–(5.43) gives

$$\begin{aligned}\left|\frac{d\alpha_{t+1}}{dx_t}\right| &\leq \frac{\bar{C}\alpha_t}{\cos\varphi_{t+1}} + E_t u_t \cos\varphi_t \left(P + \frac{Q}{\cos\varphi_t}\right) \frac{\theta^t}{\mathcal{J}_{t-1}} \\ &\leq \frac{\bar{C}\alpha_t}{\cos\varphi_{t+1}} + \frac{E_t u_t \cos\varphi_t}{R(\cos\varphi_t)} (P + C_0 Q) \frac{\theta^t}{\mathcal{J}_{t-1}}\end{aligned} \tag{5.44}$$

Since $\alpha_t < C_1 \theta^t$ for some $C_1 > 0$ and $\theta < 1$, by Lemma 5.4, the first term in (5.44) can be bounded as

$$\frac{\bar{C}\alpha_t}{\cos\varphi_{t+1}} \leq \frac{E_t u_t \cos\varphi_t}{R(\cos\varphi_t)} \frac{C_0 C_1 \bar{C} \theta^{t+1}}{\theta E_{\min} u_{\min} \cos\varphi_{t+1}}$$

We can choose θ in Lemma 5.4 so that $\theta^2 > \vartheta$, then $\theta^2 > 1/\mathcal{J}_{t-1}$. Now we select P and Q so that

$$\frac{C_0 C_1 \bar{C}}{\theta E_{\min} u_{\min}} < Q, \quad \text{and} \quad (P + C_0 Q)\theta < P$$

which completes the proof of (5.43) by induction. Now (5.33) follows from (5.43) due to (4.17), and hence Lemma 5.6 is proven. □

Remark. For a future reference, we record a bound, similar to (5.36), on the second derivative of α_t taken along the corresponding fiber of $\mathcal{G}^u_{n-k+t-1}$:

$$
(5.45) \qquad \left|\frac{d^2\alpha_t}{dx_{t-1}^2}\right| \leq \frac{\operatorname{Const} \theta^t}{\cos^3 \varphi_t} \leq \frac{\operatorname{Const} \theta^t \mathcal{J}_{t-1}}{|\gamma_{n-k+t}(x_t)|^{4/3}},
$$

here the second inequality follows from the first due to (4.17) and because $\mathcal{J}_{t-1} = \mathcal{O}(1/\cos\varphi_t)$. For $t=0$, the bound (5.45) reduces to (5.36), and for $t \geq 1$ one can use an inductive argument as above, we leave the details out.

Our bounds (5.30) and (5.31) in Lemma 5.6 are too weak for small values of $k-j$. The next lemma provides stronger bounds for that case:

LEMMA 5.7. *For some constant $\theta < 1$, we have*

$$
(5.46) \qquad \left|\int_\Omega \alpha (\ln \rho)' A_- B_+ \, d\mu\right| \leq \operatorname{Const} \theta^j
$$

$$
(5.47) \qquad \left|\int_\Omega \alpha' A_- B_+ \, d\mu\right| \leq \operatorname{Const} \theta^j
$$

The proof is based on a more general lemma, which will be useful later as well:

LEMMA 5.8. *Let $\mathcal{G}_* = \{\ell\}$ be a family of standard pairs $\ell = (\gamma_\ell, \rho_\ell)$ and λ_* a probability measure on $\mathcal{G}_*$ such that*

$$
(5.48) \qquad \lambda_*\{\ell \colon |\gamma_\ell| < \varepsilon\} \leq \operatorname{Const} \varepsilon \qquad \forall \varepsilon > 0.
$$

Let A be a C^1 function on Ω such that $\int A\, d\mu = 0$. Let $B_\ell \colon \gamma_\ell \to \mathbb{R}$ be a family of functions such that

$$
(5.49) \qquad \|B_\ell\|_\infty < b|\gamma_\ell|^{-\beta}
$$

for some $\beta \in (0,1)$ and $b > 0$, and for every ℓ and any $x, y \in \gamma_\ell$

$$
(5.50) \qquad |B_\ell(x) - B_\ell(y)| \leq b|\gamma_\ell|^{-\beta} [\operatorname{dist}(x,y)]^\varsigma
$$

for some $\varsigma > 0$. Then for some $\theta \in (0,1)$ we have

$$
\left|\int_{\mathcal{G}_*} d\lambda_* \int_{\gamma_\ell} (A \circ \mathcal{F}^n) B_\ell \rho_\ell \, dx\right| \leq \operatorname{Const} b\,\theta^n
$$

for all $n \geq 0$.

Proof. Let $k = n/2$ and
$$\mathcal{G}_*^0 = \{\ell \in \mathcal{G}_* : |\gamma_\ell| < e^{-k/K}\}$$
where $K > 0$ is the constant from Proposition A.2. Observe that
$$\left| \int_{\mathcal{G}_*^0} d\lambda_* \int_\gamma (A \circ \mathcal{F}^n) B_\ell \, \rho_\ell \, dx \right| \leq \text{Const}\, b \int_0^{e^{-k/K}} t^{-\beta} \, dt \leq \text{Const}\, b\, \theta^n$$
for some $\theta < 1$. Then we apply Proposition A.2 to every pair $(\gamma_\ell, \rho_\ell) \in \mathcal{G}_*^1 := \mathcal{G}_* \setminus \mathcal{G}_*^0$ in the following way. Denote by $\gamma_1', \gamma_2', \ldots$ the H-components of $\mathcal{F}^k(\gamma_\ell)$. On each curve $\gamma_j'' = \mathcal{F}^{-k}(\gamma_j') \subset \gamma_\ell$, we pick a point $x_j \in \gamma_j''$ and replace $B_\ell(x)$ with a constant function $\bar{B}_\ell(x) = B_\ell(x_j)$ on the curve γ_j''. This replacement gives us an error term
$$\int_{\mathcal{G}_*^1} d\lambda_* \int_\gamma |A \circ \mathcal{F}^n| \, |B_\ell - \bar{B}_\ell| \, \rho_\ell \, dx \leq \text{Const}\, \|A\|_\infty \, b\, \vartheta^{k\varsigma} \int_0^1 t^{-\beta}\, dt$$
$$\leq \text{Const}\, b\, \vartheta^{n\varsigma/2}$$
here $\vartheta < 1$ is the minimal contraction factor of u-curves under $\mathcal{F}^{-1}$. Lastly, the constant function $\bar{B}_\ell$ can be factored out, and we can apply Proposition A.2 to the H-components of $\mathcal{F}^k(\gamma_\ell)$ and get
$$\left| \int_{\mathcal{G}_*^1} d\lambda_* \int_\gamma (A \circ \mathcal{F}^n) \bar{B}_\ell \, \rho_\ell \, dx \right| \leq \text{Const}\, b\, \theta^k \int_0^1 t^{-\beta}\, dt$$
$$\leq \text{Const}\, b\, \theta^{n/2} \qquad \square$$

Remark. In the above lemma, it is obviously enough to require (5.50) only for $x, y \in \gamma_\ell$ such that $\mathcal{F}^k(x)$ and $\mathcal{F}^k(y)$ belong to the same H-component of $\mathcal{F}^k(\gamma_\ell)$. In fact, the requirements of the lemma can be relaxed even further in the following way: (5.49) may be replaced by

(5.51) $\qquad |B_\ell(x)| < \text{Const}\left(|\gamma_\ell|^{-\beta} + |\gamma_{\ell,x}'|^{-\beta}\right)$

where $\gamma_{\ell,x}'$ denotes the H-component of $\mathcal{F}(\gamma_\ell)$ that contains the point $\mathcal{F}(x)$, and (5.50) may be replaced by

(5.52) $|B_\ell(x) - B_\ell(y)| \leq \text{Const}\left(\dfrac{[\text{dist}(x,y)]^\varsigma}{|\gamma_\ell|^\beta} + \dfrac{[\text{dist}(\mathcal{F}(x), \mathcal{F}(y))]^\varsigma}{|\gamma_{\ell,x}'|^\beta}\right)$

for every $x, y \in \gamma_\ell$ such that $\mathcal{F}(y) \in \gamma_{\ell,x}'$. The proof only requires minor changes that we leave to the reader.

Proof of Lemma 5.7. It suffices to apply Lemma 5.8 to two functions, $B_{1,\ell} = \alpha(\ln \rho)' A_-$ and $B_{2,\ell} = \alpha' A_-$. The family $\mathcal{G}_*$ consists of fibers of the foliation $\mathcal{G}_{n-j-1}^u$, and (5.48) follows from the growth lemma 4.10. Next, (5.51) for the functions $B_{1,\ell}$ and $B_{2,\ell}$ follows from (5.32) and

(5.33), respectively. To verify (5.52), it is enough to show that for $r = 1, 2$

$$(5.53) \qquad |B'_{r,\ell}(x)| \leq \text{Const}\left(|\gamma_\ell|^{-q} + |\gamma'_{\ell,x}|^{-q}\mathcal{J}_{\gamma_\ell}\mathcal{F}(x)\right)$$

with some $q < 2$ (here $\mathcal{J}_{\gamma_\ell}\mathcal{F}(x)$ stands for the Jacobian of the map $\mathcal{F}\colon \gamma_\ell \to \gamma'_{\ell,x}$ at the point x). Indeed, if (5.53) holds, then for any $x, y \in \gamma_\ell$

$$|B_{r,\ell}(x) - B_{r,\ell}(y)| \leq \frac{\text{dist}(x,y)}{|\gamma_\ell|^q} + \frac{\text{dist}(\mathcal{F}x, \mathcal{F}y)}{|\gamma'_{\ell,x}|^q}$$

$$\leq \frac{[\text{dist}(x,y)]^{1-q/2}}{|\gamma_\ell|^{q/2}} + \frac{[\text{dist}(\mathcal{F}x, \mathcal{F}y)]^{1-q/2}}{|\gamma'_{\ell,x}|^{q/2}}$$

and we get (5.52). It remains to prove (5.53) for both functions $B_{1,\ell}$ and $B_{2,\ell}$. This is a consequence of the following obvious facts: $|A'_-| \leq \text{Const}$, $\alpha \leq \text{Const}$, and $\alpha' \leq \text{Const}\,|\gamma'_{\ell,x}|^{-2/3}$ by (5.33),

$$|\alpha''| \leq \text{Const}\,|\gamma'_{\ell,x}|^{-4/3}\mathcal{J}_{\gamma_\ell}\mathcal{F}(x)$$

by (5.45), $|(\ln \rho)'| \leq \text{Const}\,|\gamma_\ell|^{-2/3}$ by (B.7) and $|(\ln \rho)''| \leq \text{Const}\,|\gamma_\ell|^{-4/3}$ by (B.8). Lemma 5.7 is now proved. $\square$

Combining Lemmas 5.6 and 5.7 gives the following upper bound on all non-boundary terms in the integral formula (5.26):

COROLLARY 5.9.

$$\sum_{n=1}^{N}\sum_{k=0}^{n-1}\sum_{j=0}^{k}\left|I^{(v)}_{n,k,j}\right| \leq \text{Const}\,N$$

It remains to estimate the boundary terms $I^{(d)}_n$ and $I^{(b)}_{n,k,j}$.

5.5. Cancellation of large boundary terms. Here we estimate the boundary terms $I^{(d)}_n$ given by (5.11) and $I^{(b)}_{n,k,j}$, see (5.27). First we rewrite them in a more explicit manner and cancel out some of the resulting integrals.

Convention. Let $S \subset \Omega$ be a smooth curve, C a function and $\mathbf{v}$ a vector field on S. Then we can integrate

$$(5.54) \qquad \int_S C\,(\omega * \mathbf{v}) = \int_S C\,\|\mathbf{v}^\perp\|_0 \cos\varphi\,dl$$

were ω denotes the $\mathcal{F}$-invariant volume form

$$\omega(dr, d\varphi) = \cos\varphi\,dr \wedge d\varphi$$

and $(\omega * \mathbf{v})$ stands for the one form
$$(\omega * \mathbf{v})(\mathbf{w}) = \omega(\mathbf{v}, \mathbf{w})$$

On the right hand side of (5.54), $\|\cdot\|_0$ stands for the Euclidean norm $[(dr)^2 + (d\varphi)^2]^{1/2}$ and $\mathbf{v}^\perp$ for the normal component of the vector $\mathbf{v}$, and we integrate with respect to the Lebesgue measure (length) dl on S.

The $\mathcal{F}$-invariance of ω gives us a *change of variables formula*

$$(5.55) \qquad \int_S C\,(\omega * \mathbf{v}) = \int_{\mathcal{F}^n(S)} \left(C \circ \mathcal{F}^{-n}\right) (\omega * d\mathcal{F}^n \mathbf{v})$$

provided $\mathcal{F}^n$ is smooth on S.

First we consider $I_n^{(d)}$ given by (5.11). Each discontinuity curve $S \subset \mathcal{S}_n \setminus \mathcal{S}_0$ has the form $S = \mathcal{F}^{-k} S^+$, where $0 \leq k \leq n-1$ and $S^+ \subset \mathcal{S}_1 \setminus \mathcal{S}_0$ is a discontinuity curve for $\mathcal{F}$. Thus S changes with velocity

$$v = d\mathcal{F}^{-k} v_0 - \sum_{m=0}^{k-1} d\mathcal{F}^{-(k-m)}(X)$$

where v_0 is the speed of S^+ as it changes with Q (in the normal direction). Therefore,

$$I_n^{(d)} = I_n^{(v)} - I_n^{(x)}$$
$$= \sum_{k=0}^{n-1} I_{n,k}^{(v)} - \sum_{k=0}^{n-1} \sum_{m=0}^{k-1} I_{n,k,m}^{(x)}$$

where

$$(5.56) \qquad I_{n,k}^{(v)} = \sum_{S^+} \int_{S^+} A_{-k} \Delta B_{n-k} \, (\omega * v_0)$$

and

$$(5.57) \qquad I_{n,k,m}^{(x)} = \sum_{S^+} \int_{S^+} A_{-k} \Delta B_{n-k} \, (\omega * d\mathcal{F}^m(X))$$

(here the summation is performed over all smooth curves $S^+ \subset \mathcal{S}_1 \setminus \mathcal{S}_0$).

Furthermore, by (5.22) we have

$$d\mathcal{F}^m(X) = \sum_{j=0}^{m} \alpha_{k-m,m-j}^{(j)} + \beta_{k-m,m}$$

5. REGULARITY OF THE DIFFUSION MATRIX

Reindexing our formula by $r = k - m$, $s = m - j$, and $t = j$ gives

$$I_n^{(x)} = \sum_{\substack{r,s,t \geq 0 \\ r+s+t<n}} \sum_{S^+} \int_{S^+} A_{-(r+s+t)} \Delta B_{n-(r+s+t)} \left(\omega * \alpha_{r,s}^{(t)}\right)$$

(5.58)
$$+ \sum_{\substack{r,s \geq 0 \\ r+s<n}} \sum_{S^+} \int_{S^+} A_{-(r+s)} \Delta B_{n-(r+s)} \left(\omega * \beta_{r,s}\right)$$

The first sum contains exponentially growing (with t) integrals, but they will be cancelled shortly. At this moment we estimate the total contribution of the second sum

$$T_N := \sum_{n=1}^{N} \sum_{\substack{r,s \geq 0 \\ r+s<n}} \sum_{S^+} \int_{S^+} A_{-(r+s)} \Delta B_{n-(r+s)} \left(\omega * \beta_{r,s}\right)$$

(5.59)
$$= \sum_{\substack{r,s \geq 0 \\ r+s<N}} \sum_{S^+} \int_{S^+} A_{-(r+s)} \left[\sum_{n=r+s+1}^{N} \Delta B_{n-(r+s)}\right] \left(\omega * \beta_{r,s}\right)$$

LEMMA 5.10. *We have*

$$|T_N| \leq \text{Const } N$$

Proof. Observe that for any point $x \in S^+ \setminus (\cup_{k \geq 2} S_k)$ we have

$$\left(B_{n-(r+s)}(x)\right)_+ = \left(B_{n\pm 1-(r+s)}(x)\right)_-$$

where $(\cdot)_+$ and $(\cdot)_-$ denote the one-sided limit values of the corresponding functions, and the choice of the sign (in ± 1) in the subscript depends on the orientation of the curve S^+. Since $\Delta(B) = (B)_+ - (B)_-$ for any function B, the sum in the bracket in (5.59) telescopes, hence

$$|T_N| \leq \text{Const} \sum_{\substack{r,s \geq 0 \\ r+s<N}} \|\beta_{r,s}\|_0$$

Recall that the $\|\cdot\|_0$ norm is equivalent to $\|\cdot\|$ (Proposition 4.4) and $\|\beta_{r,s}\| \leq \text{Const } \theta^s$ (Lemma 5.4), hence $|T_N| \leq \text{Const } N$. $\square$

We now turn to $I_{n,k,j}^{(b)}$ from (5.27). The set $(\mathcal{S}_{-(n-j-1)} \cup \mathcal{S}_{j+1}) \setminus \mathcal{S}_0$ consists of s-curves $S \subset \mathcal{F}^{-m}(\mathcal{S}_1 \setminus \mathcal{S}_0)$, $0 \leq m \leq j$ and u-curves $S \subset \mathcal{F}^m(\mathcal{S}_{-1} \setminus \mathcal{S}_0)$, $0 \leq m \leq n - j - 2$. Accordingly,

$$I_{n,k,j}^{(b)} = I_{n,k,j}^{(bs)} + I_{n,k,j}^{(bu)}$$

where (using change of variables)

$$(5.60) \quad I^{(bs)}_{n,k,j} = \sum_{m=0}^{j} \sum_{S^+} \int_{S^+} A_{-(n-j-1+m)} \Delta B_{j+1-m} \left(\omega * \alpha^{(m-1)}_{n-k,k-j} \right)$$

and

$$(5.61) \quad I^{(bu)}_{n,k,j} = \sum_{m=0}^{n-j-2} \sum_{S^-} \int_{S^-} \Delta \left[A_{-(n-j-1-m)} \left(\omega * \alpha^{(-m-1)}_{n-k,k-j} \right) \right] B_{j+1+m}$$

(here the summation is performed over all the discontinuity curves $S^- \subset \mathcal{S}_{-1} \setminus \mathcal{S}_0$ of the map $\mathcal{F}^{-1}$).

First we analyze (5.60). The case $m = 0$ is special, and we combine all the terms with $m = 0$ in a separate expression:

$$(5.62) \quad I^{(bs,0)}_n = \sum_{k=0}^{n-1} \sum_{j=0}^{k} \sum_{S^+} \int_{S^+} A_{-(n-j-1)} \Delta B_{j+1} \left(\omega * \alpha^{(-1)}_{n-k,k-j} \right)$$

To deal with the other terms ($m > 0$) in (5.60), we change our indexing system to $r = n - k$, $s = k - j$, and $t = m - 1$, and obtain a total of

$$\sum_{\substack{r,s,t \geq 0 \\ r+s+t < n}} \sum_{S^+} \int_{S^+} A_{-(r+s+t)} \Delta B_{n-(r+s+t)} \left(\omega * \alpha^{(t)}_{r,s} \right)$$

which completely cancels the first sum in (5.58), hence all the large integrals are now gone.

Next we make a general remark. Every curve S^- separates two regions, one is mapped by $\mathcal{F}^{-1}$ into a vicinity of $\mathcal{S}_0$ and the other – into a vicinity of some curve $S^+ \subset \mathcal{S}_1 \setminus \mathcal{S}_0$. On the side of S^- that is mapped onto $\mathcal{S}_0$, the map $\mathcal{F}^{-1}$ has unbounded derivatives, and we call that side of S^- *irregular*. On the other side, the map $\mathcal{F}^{-1}$ has bounded derivatives, and we call that side of S^- *regular*. Thus, every curve $S^- \subset \mathcal{S}_{-1} \setminus \mathcal{S}_0$ has one regular side and one irregular side. Similarly we define regular and irregular sides for every curve $S^+ \subset \mathcal{S}_1 \setminus \mathcal{S}_0$. Note that $\mathcal{F}^{-1}$ maps the regular sides of $\mathcal{S}_{-1} \setminus \mathcal{S}_0$ to the regular sides of $\mathcal{S}_1 \setminus \mathcal{S}_0$, and the map $\mathcal{F}$ does the opposite.

Observe that the integrand in (5.61) vanishes on the irregular side of every curve S^- (to see this, note that the field $\alpha^{(-m-1)}_{n-k,k-j} \subset E^u_{n-j-1-m}$ is in fact tangent to S^- on its irregular side; or, equivalently, one can approximate S^- by a curve S lying on its irregular side, apply (5.55) with $n = -1$ to S, and note that the form ω vanishes on $\mathcal{F}^{-1}(S)$ as that curve approaches $\mathcal{S}_0$). Now we can change variables $y = \mathcal{F}^{-1}x$

5. REGULARITY OF THE DIFFUSION MATRIX

and rewrite (5.61) as

$$(5.63) \quad I^{(bu)}_{n,k,j} = \sum_{m=1}^{n-j-1} \sum_{S^+} \int_{S^+} A_{-(n-j-1-m)} B_{j+1+m} \left(\omega * \alpha^{(-m-1)}_{n-k,k-j} \right)$$

where the integration is performed along the regular side of each curve S^+ (again, on the irregular side of S^+ the integrand in (5.63) vanishes).

Now the integrals in (5.63) can be naturally combined with those in (5.62) and make a total of

$$(5.64) \quad I^{(b)}_{n,k} = \sum_{j=0}^{k} \sum_{m=0}^{n-j-1} \sum_{S^+} \int_{S^+} A_{-(n-j-1-m)} B_{j+1+m} \left(\omega * \alpha^{(-m-1)}_{n-k,k-j} \right)$$

Here the case $m = 0$ corresponds to (5.62) and the case $m \geq 1$ to (5.63). (Note that the integrand in (5.62) also vanishes on the irregular side of each curve S^+.)

We also note that

$$(5.65) \quad \left\| \alpha^{(-m-1)}_{n-k,k-j} \right\| \leq \text{Const}\, \theta^{m+k-j}$$

due to Lemma 5.4.

It remains to estimate the terms $I^{(v)}_{n,k}$ given by (5.56) and $I^{(b)}_{n,k}$ of (5.64).

Remark. Before proceeding with our estimates let us compare the approach of the present section with that of Chapter 4. There are three types of terms corresponding to the variation of $\mu(A(B \circ \mathcal{F}_Q))$:

- "stable continuous" terms $I^s_{n,k}$ given by (5.24). They correspond to the term $I\!I$ in (4.42) since they deal with the difference of the values of the observable at the shadowed and the shadowing points.
- "unstable continuous" terms $I^{(i)}_{n,k,j}$ given by (5.28). They correspond to the term $I\!I\!I$ in (4.42), since the Jacobian of the holonomy map is a product of unstable Jacobian ratios.
- The terms containing integration over discontinuity (T_N given by (5.59), $I^{(v)}_{n,k}$ given by (5.56) and $I^{(b)}_{n,k}$ given by (5.64)). They correspond to the first term in (4.42) since they account for orbits where shadowing is impossible as they pass too close to the singularities.

5.6. Estimation of small boundary terms. First we outline our strategy. All the integrals in (5.56) and (5.64) have a general form of

$$\int_{S^+} A_{-k_1} B_{k_2} (\omega * \mathbf{v}) = \int_{S^+} A_{-k_1} B_{k_2} \|\mathbf{v}^\perp\|_0 \cos \varphi \, dl$$

with $k_1 + k_2 = n$ and some vector fields $\mathbf{v}$ on S^+. The curve S^+ is strongly expanded by $\mathcal{F}^{-k_1}$, as well as by $\mathcal{F}^{k_2}$, and so both functions A_{-k_1} and B_{k_2} rapidly oscillate on the curve S^+. However, if $k_1 \ll k_2$, then B_{k_2} oscillates much faster than A_{-k_1}, and we will approximate A_{-k_1} by constants on appropriately chosen pieces of S^+ and then use Proposition A.2 to average B_{k_2} on each of those pieces. If $k_1 \gg k_2$, then A_{-k_1} and B_{k_2} switch places. In the remaining case $k_1 \approx k_2$ we simply bound the above integrand by $\|A\|_\infty \|B\|_\infty \sup_\Omega \|\mathbf{v}\|$, and then summing up over $n \leq N$ and using (5.65) will give us the desired $\mathcal{O}(N)$ estimate.

When applying Proposition A.2, we will treat the function $\rho = \|\mathbf{v}^\perp\|_0 \cos\varphi$ as a "density" on the corresponding pieces of S^+, so that they become standard pairs. However, while v_0 in (5.56) is bounded and smooth (which can be easily verified directly, we omit details), the vector fields (and hence, the corresponding ρ) in (5.64) are badly discontinuous: their discontinuities lie on the set $\mathcal{S}_{-k_1}$, which is very dense on S^+. Our first task is to approximate vector fields in (5.64) by smooth enough functions. To this end we develop a general approach.

Let $S \subset \Omega$ be a u-curve or an s-curve, $a_1 \in (0,1]$ and $a_2 \geq 0$. We denote by $\mathcal{H}^{a_1,a_2}(S)$ the class of functions $\rho\colon S \to \mathbb{R}$ that are well approximated by Hölder continuous functions in the following sense:

Definition. $\rho \in \mathcal{H}^{a_1,a_2}(S)$ iff there is a $L_\rho > 0$ such that for every $\varepsilon \in (0,1)$ there exists a function $\rho_\varepsilon\colon S \to \mathbb{R}$ satisfying two requirements:

(5.66)
$$\int_S |\rho - \rho_\varepsilon|\, dl \leq \varepsilon$$

and for all $x, y \in S$

(5.67)
$$|\rho_\varepsilon(x) - \rho_\varepsilon(y)| \leq L_\rho \varepsilon^{-a_2} |S(x,y)|^{a_1}$$

where $S(x,y)$ denotes the segment of the curve S between the points x and y. We always take the smallest L_ρ for which (5.67) holds for all $\varepsilon \in (0,1)$ and put
$$\|\rho\|_{a_1,a_2} := L_\rho.$$

LEMMA 5.11 (Hölder approximation). *There exist* $a_1 \in (0,1]$ *and* $a_2 \geq 0$ *such that*
$$\rho = \left\|\left(\alpha_{n-k,k-j}^{(-m-1)}\right)^\perp\right\|_0 \cos\varphi \in \mathcal{H}^{a_1,a_2}(S^+)$$

and
$$\|\rho\|_{a_1,a_2} \leq 1$$

uniformly in n, k, j, m.

We postpone the proof of Lemma 5.11 until Section 5.9 and continue our analysis of the integrals (5.56) and (5.64).

As we mentioned already, v_0 is a bounded and smooth vector field, hence

(5.68) $\qquad \|v_0\|_\infty \leq \text{Const} \qquad$ and $\qquad \|v_0\|_{a_1,a_2} \leq \text{Const}$

for any $a_1 \in (0,1]$ and $a_2 \geq 0$.

PROPOSITION 5.12 (Two-sided integral sums). *Given $a_1 \in (0,1]$, $a_2 \geq 0$, and $L > 0$, there are constants $C, c, \xi > 0$ such that for each curve $S^+ \subset S_1 \setminus S_0$ and all m_1, m_2 such that $m_j < L \ln N$, for any $\delta > 0$, and for any functions $\rho_{k_1} \in \mathcal{H}^{a_1,a_2}(S^+)$ such that*

(5.69) $\qquad \|\rho_{k_1}\|_\infty \leq \delta \qquad$ and $\qquad \|\rho_{k_1}\|_{a_1,a_2} \leq 1$

we have

$$\left| \sum_{\substack{k_1 > m_1, k_2 > m_2 \\ k_1 + k_2 \leq N}} \int_{S^+} A_{-k_1} B_{k_2} \rho_{k_1} \, dl \right| \leq C \left(N\delta |\ln \delta| + N^2 e^{-cN^\xi} \right)$$

where the integral can be taken on either side of S^+ (but this should be done consistently).

We prove Proposition 5.12 in Section 5.7.

COROLLARY 5.13.

$$\left| \sum_{n=1}^{N} \sum_{k=0}^{n-1} I_{n,k}^{(v)} \right| \leq \text{Const } N, \qquad \left| \sum_{n=1}^{N} \sum_{k=0}^{n-1} I_{n,k}^{(b)} \right| \leq \text{Const } N$$

Proof. We prove the second bound (the first one is easier). Introduce new indices (k_1, k_2, r) where $k_1 = n - j - 1 - m$, $k_2 = j + 1 + m$, and $r = k - j$. Due to (5.65) we can choose L so large that the sum over quadruples with $m > L \ln N$ or $r > L \ln N$ will be less than 1. Now Proposition 5.12 and Lemma 5.4 imply that for fixed m and r such that $m \leq L \ln N$ and $r \leq L \ln N$, the sum over k_1 and k_2 is bounded by $\text{Const}\,[(r+m)\theta^{r+m} N + N^2 e^{-cN^\xi}]$. Summation over m and r gives the desired bound. $\square$

This completes the proof of our main estimate (5.10) and hence that of Proposition 5.2 (modulo Lemma 5.11 and Proposition 5.12). $\square$

5.7. Two-sided integral sums. Here we prove Proposition 5.12. For the sake of brevity we shall call any set of the form
$$\{(k_1, k_2): k_1 \geq m_1,\ k_2 \geq m_2,\ (k_1 - m_1) + (k_2 - m_2) \leq R\}$$
a *triangle* with side R, and any set of the form
$$\{(k_1, k_2): m_1 \leq k_1 \leq m_1 + R,\ m_2 \leq k_2 \leq m_2 + R\}$$
a *square* with side R. For brevity, we denote
$$\mathcal{I}_{k_1,k_2} = \int_{S^+} A_{-k_1} B_{k_2} \rho_{k_1}\, dl$$

LEMMA 5.14. *For any square* $\mathbf{S}_R$ *with side* R
$$\left| \sum_{(k_1,k_2) \in \mathbf{S}_R} \mathcal{I}_{k_1,k_2} \right| \leq \operatorname{Const} R\,\delta |\ln \delta|$$

Proof. For simplicity, we will set $m_1 = m_2 = 0$ (the general case only requires minor modifications). Now we have
$$\left| \sum_{(k_1,k_2) \in \mathbf{S}_R} \mathcal{I}_{k_1,k_2} \right| \leq \int_{S^+} \left| \left(\sum_{k_1} A_{-k_1} \rho_{k_1} \right) \left(\sum_{k_2} B_{k_2} \right) \right| dl$$
$$\leq \left[\left(\int_{S^+} \left[\sum_{k_1} A_{-k_1} \rho_{k_1} \right]^2 dl \right) \left(\int_{S^+} \left[\sum_{k_2} B_{k_2} \right]^2 dl \right) \right]^{1/2}$$

To estimate the first factor we expand
$$\int_{S^+} \left[\sum_{k_1} A_{-k_1} \rho_{k_1} \right]^2 dl = \int_{S^+} \sum_{j_1, j_2} A_{-j_1} A_{-j_2} \rho_{j_1} \rho_{j_2}\, dl$$
$$= \int_{S^+} \sum_{j} A_{-j}^2 \rho_j^2\, dl + 2 \int_{S^+} \sum_{j_2 > j_1} A_{-j_1} A_{-j_2} \rho_{j_1} \rho_{j_2}\, dl$$

The first term here is $\mathcal{O}(R\delta^2)$. To estimate the second sum we choose a large $K > 0$ and divide the domain of summation $\{j_1 < j_2\} \subset \mathbf{S}_R$ into two parts: a smaller one
$$\mathbf{S}'_R = \{j_1 < 2K|\ln \delta|\} \cup \{|j_1 - j_2| < 2K|\ln \delta|\}$$
and a larger one $\mathbf{S}''_R = \{j_1 < j_2\} \setminus \mathbf{S}'_R$. Obviously,
$$\left| \int_{S^+} \sum_{(j_1,j_2) \in \mathbf{S}'_R} A_{-j_1} A_{-j_2} \rho_{j_1} \rho_{j_2}\, dl \right| \leq \operatorname{Const} R\,\delta^2 |\ln \delta|$$

To estimate the larger sum, we need to approximate $\rho_{j_1,j_2} = \rho_{j_1}\rho_{j_2}$ by a Hölder continuous function: (5.69) implies $\rho_{j_1,j_2} \in \mathcal{H}^{a_1,a_2}(S^+)$ and $\|\rho_{j_1,j_2}\|_{a_1,a_2} \leq 1$, hence we can set $\varepsilon = e^{-j_1/K}$ and find $\bar{\rho}_{j_1,j_2}$ such that

$$\int_{S^+} |\rho_{j_1,j_2} - \bar{\rho}_{j_1,j_2}|\, dl \leq e^{-j_1/K}$$

and for any $x, y \in S^+$

$$|\bar{\rho}_{j_1,j_2}(x) - \bar{\rho}_{j_1,j_2}(y)| \leq e^{a_2 j_1/K}[\mathrm{dist}(x,y)]^{a_1}$$

The error of approximation can be bounded by

$$\int_{S^+} \sum_{(j_1,j_2)\in \mathbf{S}''_R} |A_{-j_1} A_{-j_2}|\,|\rho_{j_1,j_2} - \bar{\rho}_{j_1,j_2}|\, dl \leq \|A\|_\infty^2 \sum_{j_2 > j_1 \geq 2K|\ln\delta|} e^{-j_1/K}$$
$$\leq \mathrm{Const}\, R\delta^2$$

It remains to bound the integrals

$$\bar{\mathcal{I}}_{j_1,j_2} = \int_{S^+} A_{-j_1} A_{-j_2} \bar{\rho}_{j_1,j_2}\, dl$$

for $(j_1, j_2) \in \mathbf{S}''_R$. We denote by S^+_q, $q \geq 1$, all the H-components of $\mathcal{F}^{-j_1}(S^+)$ (i.e. the maximal curves $S^+_q \subset \mathcal{F}^{-j_1}(S^+)$ such that $\mathcal{F}^i(S^+_q)$ lies in one homogeneity strip for each $i = 0, \ldots, j_1$), and by m_q the image of the Lebesgue measure dl under $\mathcal{F}^{-j_1}$ on S^+_q. Then

$$(5.70) \qquad \bar{\mathcal{I}}_{j_1,j_2} = \sum_q \int_{S^+_q} (A\, A_{-(j_2-j_1)})(\bar{\rho}_{j_1,j_2} \circ \mathcal{F}^{j_1})\, dm_q$$

We claim that if $K > 0$ is large enough, then the function $g = \bar{\rho}_{j_1,j_2} \circ \mathcal{F}^{j_1}$ is Hölder continuous on each curve S^+_q with exponent a_1 and a uniformly bounded norm. Indeed, for any $x, y \in S^+_q$

$$|g(x) - g(y)| \leq e^{a_2 j_1/K} \vartheta^{a_1 j_1}[\mathrm{dist}(x,y)]^{a_1} \leq [\mathrm{dist}(x,y)]^{a_1}$$

where $\vartheta < 1$ is the minimal factor of expansion of s-curves under $\mathcal{F}^{-1}$, and $e^{a_2/K}\vartheta^{a_1} < 1$ provided K is large enough. Hence we can apply Lemma 5.8 to the map $\mathcal{F}^{-(j_2-j_1)}$ (using time reversibility) on the set $\cup_q S^+_q$, and thus estimate (5.70) as

$$\left|\sum_q \int_{S^+_q} (A\, A_{-(j_2-j_1)})(\bar{\rho}_{j_1,j_2} \circ \mathcal{F}^{j_1})\, dm_q\right| \leq \mathrm{Const}\, \theta^{j_2-j_1}$$

with some constant $\theta < 1$. Therefore

$$\sum_{(j_1,j_2)\in \mathbf{S}''_R} |\bar{\mathcal{I}}_{j_1,j_2}| \leq \mathrm{Const}\, R\, \theta^{2K|\ln\delta|} \leq \mathrm{Const}\, R\delta^2$$

provided K is large enough (say, $K > 1/|\ln\theta|$). Combining all the previous estimates gives

$$\int_{S^+} \left(\sum_{k_1} A_{-k_1} \rho_{k_1} \right)^2 dl \leq \text{Const } R\,\delta^2 |\ln\delta|$$

The same argument yields

$$\int_{S^+} \left(\sum_{k_2} B_{k_2} \right)^2 dl \leq \text{Const } R$$

(in fact, this is easier since there is no ρ's to approximate). This completes the proof of Lemma 5.14. □

LEMMA 5.15. *There exists a constant $C > 0$ such that for any triangle $\mathbf{T}$ with side R*

$$\left| \sum_{(k_1,k_2)\in \mathbf{T}} \mathcal{I}_{k_1,k_2} \right| \leq C(R\ln R)\,\delta |\ln\delta|$$

Proof. (See Figure 3.) We decompose $\mathbf{T}$ into the union of a square and two triangles with sides $R/2$. Then we apply a similar decomposition to each of the two smaller triangles, an so on. In this way we get a decomposition of $\mathbf{T}$ into squares of variable size, so that for each $k \geq 1$ there are 2^k squares with side about $R/2^k$. Applying Lemma 5.14 to each square yields the required bound. □

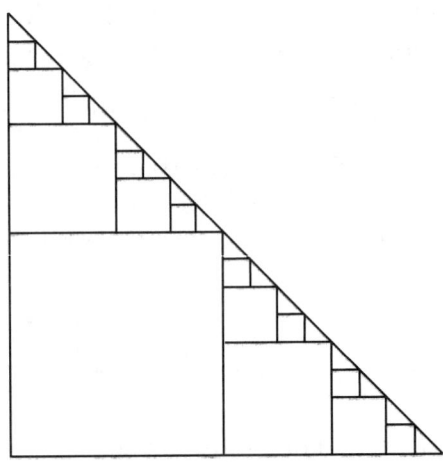

FIGURE 3. Proof of Lemma 5.15

5. REGULARITY OF THE DIFFUSION MATRIX

Lemma 5.15 falls short of the estimate claimed in Proposition 5.12, because of the extra $\ln R$ factor here, but it has the advantage of being applicable to an arbitrary triangle. To upgrade Lemma 5.15 to the estimate claimed in Proposition 5.12 we need to bound off-diagonal terms.

LEMMA 5.16 (Off-diagonal bounds). *Fix some $0 < \zeta < 1/2$. Then there are constants $C, c, \xi > 0$ such that if*

$$\max\{k_1, k_2\} > R/2 \quad \text{and} \quad |k_1 - k_2| > R^{1/2+\zeta} \tag{5.71}$$

then

$$|\mathcal{I}_{k_1, k_2}| \leq C \exp(-cR^\xi)$$

We prove Lemma 5.16 in Section 5.8 and first derive Proposition 5.12.

Proof of Proposition 5.12. (See Figure 4.) Let **T** be the triangle of Proposition 5.12, **S** the inscribed square and $\mathbf{T}_1, \mathbf{T}_2$ the triangles with side $N^{1/2+\zeta}$ whose one vertex is the midpoint of the hypotenuse of **T**. Then

$$\sum_{(k_1, k_2) \in \mathbf{T}} \mathcal{I}_{k_1, k_2} = \sum_{\mathbf{S}} \mathcal{I}_{k_1, k_2} + \sum_{\mathbf{T}_1 \cup \mathbf{T}_2} \mathcal{I}_{k_1, k_2} + \sum_{\mathbf{T} \setminus (\mathbf{S} \cup \mathbf{T}_1 \cup \mathbf{T}_2)} \mathcal{I}_{k_1, k_2}$$

The first sum here is $\mathcal{O}(N\delta|\ln \delta|)$ by Lemma 5.14, the second one is $\mathcal{O}(N^{1/2+\zeta} \ln N \delta |\ln \delta|)$ by Lemma 5.15, and the last one is $\mathcal{O}(N^2 \times \exp(-cN^\xi))$ by Lemma 5.16, because every pair $(k_1, k_2) \in \mathbf{T} \setminus (\mathbf{S} \cup \mathbf{T}_1 \cup \mathbf{T}_2)$ satisfies (5.71). $\square$

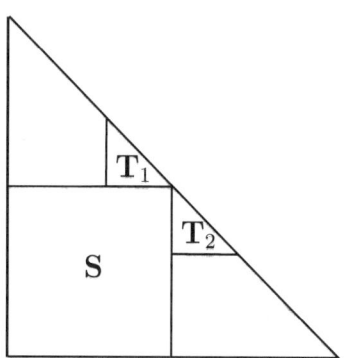

FIGURE 4. Proof of Proposition 5.12

5.8. Bounding off-diagonal terms. Here we prove Lemma 5.16. Our main idea is that if $k_2 - k_1 \gg \sqrt{R}$, then we can partition S^+ into subintervals such that the preimages under $\mathcal{F}^{-k_1}$ are predominantly small whereas their images under $\mathcal{F}^{k_2}$ are mostly large (as it will follow from moderate deviation bounds of Section A.4). Thus we can approximate A_{-k_1} and ρ_{k_1} by constants on each interval and average the value of B_{k_2} by using Proposition A.2.

Without loss of generality, we suppose that
$$k_2 - k_1 > \Delta = R^{1/2+\varsigma}$$
(the case $k_1 - k_2 > \Delta$ is completely symmetric by time-reversibility).

Let $k_3 = k_2 - \Delta/4$ and $k_4 = k_2 - \Delta$. Denote by S_p^-, $p \geq 1$, all the H-components of the set $\mathcal{F}^{k_3}(S^+)$, then the curves $S_p^+ = \mathcal{F}^{-k_3}(S_p^-) \subset S^+$ make a partition of S^+. Let χ denote the Lyapunov exponent of the map $\mathcal{F}$. We say that a curve S_p^+ is *good* if it satisfies three requirements:

(a) $|S_p^+| < \exp(-\chi k_3 + \chi\Delta/4)$,
(b) $\mathcal{F}^{-k_4}(S_p^+)$ belongs in one H-component of the set $\mathcal{F}^{-k_4}(S^+)$,
(c) $|\mathcal{F}^{-k_4}(S_p^+)| < \exp(-\chi\Delta/4)$,

and denote $G^+ = \cup\{S_p^+ : S_p^+ \text{ is good}\}$.

LEMMA 5.17.
(5.72) $$l(S^+ \setminus G^+) \leq \text{Const } \exp(-cR^{2\varsigma})$$
for some constant $c > 0$.

Proof. We note that the distortions of the map $\mathcal{F}^{k_3}$ on each curve S_p^+ are bounded, i.e. for any $x, y \in S_p^+$
$$0 < \bar{C}^{-1} \leq \frac{\mathcal{J}_{S_p^+}\mathcal{F}^{k_3}(x)}{\mathcal{J}_{S_p^+}\mathcal{F}^{k_3}(y)} \leq \bar{C} < \infty$$
where $\mathcal{J}_S\mathcal{F}^k(x)$ denotes the Jacobian (the expansion factor) of the map $\mathcal{F}^k$ restricted to the curve S at the point x. Now by Proposition A.7 on moderate deviations
$$l\big(\cup S_p^+ : |S_p^+| > \exp(-\chi k_3 + \chi\Delta/4)\big) < \text{Const } \exp(-cR^{2\varsigma})$$
for some $c > 0$, hence we may ignore the curves on which (a) fails.

Similarly, the distortions of the maps $\mathcal{F}^{-k}$, $k \geq 1$, on each curve S_p^+ remain bounded as long as the preimage $\mathcal{F}^{-k}(S_p^+)$ lies within one H-component of the set $\mathcal{F}^{-k}(S^+)$. If the condition (b) fails, then there is a (smallest) $k < k_4$ such that $\mathcal{F}^{-k}(S_p^+)$ crosses either a singularity line of the map $\mathcal{F}^{-1}$ or the boundary of a homogeneity strip. Now we distinguish two cases:

(b1) $\left|\mathcal{F}^{-k}(S_p^+)\right| < \exp(-\chi\Delta/4)$ (a *short* curve),
(b2) $\left|\mathcal{F}^{-k}(S_p^+)\right| \geq \exp(-\chi\Delta/4)$ (a *long* curve).

Observe that every short curve $\mathcal{F}^{-k}(S_p^+)$ lies within a distance less than $\exp(-\chi\Delta/4)$ of an endpoint of an H-component of $\mathcal{F}^{-k}(S^+)$. Therefore, by the growth lemma 4.10

$$l\left(\cup S_p^+ \colon \text{(b) fails and } \mathcal{F}^{-k}(S_p^+) \text{ is short}\right) < \text{Const}\, k_4 \exp(-\chi\Delta/4)$$

On the other hand, if $\mathcal{F}^{-k}(S_p^+)$ is long and (a) holds, then by bounded distortion

$$\mathcal{J}_{S_p^+}\mathcal{F}^{-k}(x) \geq \bar{C}^{-1} \exp\left(\chi k_3 - \chi\Delta/2\right)$$

for every point $x \in S_p^+$. Note that $k < k_4 = k_3 - 3\Delta/4$, hence by Proposition A.7 on moderate deviations

$$l\left(\cup S_p^+ \colon \text{(b) fails and } \mathcal{F}^{-k}(S_p^+) \text{ is long}\right) < \text{Const}\, k_4 \exp(-cR^{2\zeta})$$

with some $c > 0$, hence we may ignore the curves on which (b) fails.

Lastly, if (a) and (b) hold but (c) fails, then we can apply the previous argument to $\mathcal{F}^{-k_4}(S_p^+)$, since its length exceeds $\exp(-\chi\Delta/4)$. □

Now, consider a good curve $S_p^+ \subset S^+$. Observe that $k_4 \geq k_1$, hence $\left|\mathcal{F}^{-k_1}(S_p^+)\right| < \exp(-\chi\Delta/4)$. Therefore, the oscillations of the function A_{-k_1} on S_p^+ does not exceed $\text{Const}\exp(-\chi\Delta/4)$, so we can approximate A_{-k_1} by a constant, $\hat{A}_{-k_1}$, on each good curve.

Next, we use a Hölder continuous approximation to ρ_{k_1}. We set $\varepsilon = \exp(-\Delta)$ and find a $\bar{\rho}_{k_1}$ such that

$$\int_{S^+} |\rho_{k_1} - \bar{\rho}_{k_1}|\, dl \leq e^{-\Delta}$$

and for any $x, y \in S_p^+$

$$|\bar{\rho}_{k_1}(x) - \bar{\rho}_{k_1}(y)| \leq e^{a_2\Delta}[\text{dist}(x,y)]^{a_1} \leq e^{a_2\Delta - a_1\chi k_3}$$

so that the oscillations of $\bar{\rho}_{k_1}$ on S_p^+ do not exceed e^{-cR} with some $c > 0$. Now we approximate $\bar{\rho}_{k_1}$ by a constant, $\hat{\rho}_{k_1}$, on each good curve S_p^+. The errors of this and other approximations above are all bounded by $\exp(-c\Delta)$ with some $c > 0$.

Lastly, we apply Proposition A.2 to the H-components of the set $\mathcal{F}^{k_3}(S^+)$ and average the function $B \circ \mathcal{F}^{k_2-k_3}$ on every such component. This gives us the estimate

$$\left|\int_{G^+} \hat{A}_{-k_1} B_{k_2} \hat{\rho}_{k_1}\, dl\right| \leq \text{Const}\, \theta_0^{k_2-k_3} = \text{Const}\, \theta_0^{\Delta/4}$$

with a constant $\theta_0 < 1$. This proves Lemma 5.16. □

5.9. Hölder approximation. Here we prove Lemma 5.11. The problem we face is that the discontinuities of our vector field $\alpha_{n-k,k-j}^{(-m-1)}$ on the curve S^+ exponentially grow with $n-j$, whereas we need a bound independent of n, j, m. The discontinuities of our vector field are generated by those of the unstable foliation $\mathcal{G}_{n-j-m-1}$ obtained by iterating the original smooth foliation $\mathcal{G}$, see Section 5.3. However, the action of $d\mathcal{F}$ on the projective tangent space is contractive within the unstable cone. Since contractions improve smoothness, the influence of the singularities that have occurred far back in the past decays exponentially allowing us to get a uniform estimate in the end.

The singularity curves $S^+ \subset \mathcal{S}_1 \setminus \mathcal{S}_0$ are known to be C^2 smooth with uniformly bounded curvature (see [**19**], where this fact is proved even for a more general class of billiards, those in small external fields), hence it is enough to prove that the restriction of the function

$$\hat{\rho} = \|\alpha_{n-k,k-j}^{(-m-1)}\|_0$$

to S^+ has the required properties, i.e. $\hat{\rho} \in \mathcal{H}^{a_1,a_2}(S^+)$ and $\|\hat{\rho}\|_{a_1,a_2} \leq$ Const. Next, it suffices to construct approximating functions $\hat{\rho}_\varepsilon$ for

(5.73) $$\varepsilon \leq \varepsilon_0 := \|\hat{\rho}\|_\infty^d$$

for some fixed large d. Indeed, if we can construct $\hat{\rho}_\varepsilon$ satisfying (5.66)–(5.67) for some a_1, a_2 and all $\varepsilon \leq \varepsilon_0$, then for $\varepsilon > \varepsilon_0$ we can define

$$\hat{\rho}_\varepsilon = \begin{cases} 0 & \text{if } \varepsilon \geq \|\hat{\rho}\|_\infty, \\ \hat{\rho}_{\varepsilon_0} & \text{if } \|\hat{\rho}\|_\infty > \varepsilon > \varepsilon_0, \end{cases}$$

and the resulting family $\{\hat{\rho}_\varepsilon\}$ will satisfy (5.66)–(5.67) for all $\varepsilon > 0$ but with a different a_2. (In fact, the method we present here works also for $\varepsilon \geq \varepsilon_0$, but we consider only small ε in order to avoid dealing with too many different cases.)

Now let

(5.74) $$r = K|\ln \varepsilon|$$

where $K > 0$ is a sufficiently large constant. Denote by S_p^+, $p \geq 1$, all the H-components of $\mathcal{F}^{-r}(S^+)$ and let ξ^+ be the partition of S^+ into the curves $\mathcal{F}^r(S_p^+)$. Next, $\mathcal{F}$ maps S^+ onto a u-curve $S^- \subset \mathcal{S}_{-1} \setminus \mathcal{S}_0$, and we denote by S_q^-, $q \geq 1$, all the H-components of $\mathcal{F}^{r-1}(S^-)$. Denote by ξ^- the partition of S^+ into the curves $\mathcal{F}^{-r}(S_q^-)$. Let $\xi = \xi^+ \vee \xi^-$ and denote by $\xi(x)$ the element of the partition ξ that contains the point x.

We say that an element $W \in \xi$ is *large* if length$(W) > \varepsilon^{K^2}$ and *small* otherwise. We claim that

5. REGULARITY OF THE DIFFUSION MATRIX

LEMMA 5.18. *The total Lebesgue measure of small intervals is less than* $\mathrm{Const}\, \varepsilon^2$ *if K is large enough.*

Proof. It is enough to check that
$$l\{x\colon \mathrm{dist}(x, \partial \xi(x)) < \varepsilon^{K^2}\} < \mathrm{Const}\, \varepsilon^2$$
This in turn follows from the estimates
$$l\{x\colon \mathrm{dist}(x, \partial \xi^{\pm}(x)) < \varepsilon^{K^2}\} < \mathrm{Const}\, \varepsilon^2$$
and the last bound holds by Proposition A.5 on large deviations and the growth lemma 4.10. □

Next, we consider an element $W \in \xi$. Observe that (5.73), (5.74) and (5.65) imply $r \gg k - j + m$ (in fact, even $r \gg 2(k - j + m)$), so that $n - k \gg n - j - m - r$. Hence, the projectors Θ^s_p and Θ^u_p used in the construction of the field $\alpha^{(-m-1)}_{n-k,k-j}$, cf. Section 5.3, are smooth on the curve W and its preimages, up to $\mathcal{F}^{-(n-k)}(W)$. Hence all the discontinuities of $\alpha^{(-m-1)}_{n-k,k-j}$ on W come from the discontinuities of the field $E^u_{n-j-m-1-r}$ on the curve $\mathcal{F}^{-r}(W)$.

We claim that there exist constants $\tilde{\theta} < 1$ and $a_1 > 0$ such that the restriction of $\alpha^{(-m-1)}_{n-k,k-j}$ to each $W \in \xi$ can be $\tilde{\theta}^r$ approximated in the L^∞ metric by a vector field whose a_1-Hölder norm is uniformly bounded. To prove this claim, consider first the case $r \leq n - j - 1 - m$. We take an arbitrary smooth family of unstable directions $\hat{E}^u$ on the curve $\mathcal{F}^{-r}(W)$ (whose derivative along $\mathcal{F}^{-r}(W)$ is uniformly bounded), for example, we take the restriction of the family E^u defined in Section 5.3 to $\mathcal{F}^{-r}(W)$. Then we use $\hat{E}^u$, instead of $E^u_{n-j-1-m-r}$, to construct an approximation $\hat{\alpha}_{n,k,j,m,r}$ to $\alpha^{(-m-1)}_{n-k,k-j}$. Our claim now follows from a general fact:

Fact. Given a smooth field $\hat{E}^u$ of unstable directions, the derivatives of $\mathcal{F}^n \hat{E}^u$ along stable curves grow exponentially with n, but the field $\mathcal{F}^n \hat{E}^u$ remains Hölder continuous with some fixed exponent $a > 0$ and a Hölder norm bounded uniformly over all $n \geq 0$.

This fact is known as the Invariant Section Theorem [**80**, Theorem 5.18], and it has been proven for quite general hyperbolic systems. For dispersing billiards, we outline a direct proof in Section B.2.

Hence we obtain an a_1-Hölder continuous vector field $\hat{\alpha}_{n,k,j,m,r}$ with a uniformly bounded Hölder norm on each W. Besides,
$$\left\| \hat{\alpha}_{n,k,j,m,r} - \alpha^{(-m-1)}_{n-k,k-j} \right\|_\infty \leq \mathrm{Const}\, \tilde{\theta}^r$$

5. REGULARITY OF THE DIFFUSION MATRIX

because the angle between $\mathcal{F}^r(\hat{E}^u)$ and $E^u_{n-j-1-m}$ at every point $x \in S^+$ is $\mathcal{O}(\tilde{\theta}^r)$.

This proves the claim in the case $r \leq n-j-1-m$. If the opposite inequality holds, then the field $\alpha^{(-m-1)}_{n-k,k-j}$ itself is smooth on W and the claim follows by a direct application of the Invariant Section Theorem.

Lastly, we need to make our approximative vector field Hölder continuous on the entire curve S^+, which will be done in two steps. First, let $\hat{\alpha}^{(1)}_{n,k,j,m,r}$ coincide with $\hat{\alpha}_{n,k,j,m,r}$ on large intervals $W \subset S^+$ and be 0 on small ones (see the definition above). We have

$$\int_{S^+} \|\hat{\alpha}^{(1)}_{n,k,j,m,r} - \alpha^{(-m-1)}_{n-k,k-j}\|_0 \, dl \leq \text{Const}(\varepsilon^2 \theta^{k-j+m} + \tilde{\theta}^r)$$

where the first term estimates the contribution of the small intervals via Lemma 5.18 and (5.65) and the second term estimates the contribution of the large intervals via (5.9). This modification eliminates most of the discontinuities of $\hat{\alpha}_{n,k,j,m,r}$, however $\hat{\alpha}^{(1)}_{n,k,j,m,r}$ is not yet globally Hölder continuous – it can have jumps of size $\mathcal{O}(\theta^{m+k-j})$ at the endpoints of each large interval. The total number of jumps is twice the number of large intervals, i.e. $\leq \text{Const}\,\varepsilon^{-K^2}$. Now we further modify $\hat{\alpha}^{(1)}_{n,k,j,m,r}$ by replacing it with a linear function in the ε^{K^2+1} neighborhood of each jump, so that the new modification, we call it $\hat{\alpha}^{(2)}_{n,k,j,m,r}$, becomes continuous on S^+. It is easy to see that

$$\int_{S^+} \|\hat{\alpha}^{(2)}_{n,k,j,m,r} - \alpha^{(-m-1)}_{n-k,k-j}\|_0 \, dl \leq \text{Const}(\varepsilon \theta^{k-j+m} + \tilde{\theta}^r)$$

and the a_1-Hölder norm of the new approximation is $\leq \text{Const}\,\varepsilon^{-a_1(K^2+1)}$. So we set $a_2 = a_1(K^2+1)$ and obtain the required approximation to $\alpha^{(-m-1)}_{n-k,k-j}$ on S^+. □

CHAPTER 6

Moment estimates

The main goal of this section is to prove Proposition 3.5 and thus establish Theorem 2. Our proof is based on various moment estimates of the underlying processes. In addition, we obtain some more estimates to be used in the proofs of Theorems 1 and 3 presented in the subsequent sections.

6.1. General plan. Here we formulate several propositions that constitute the basis of our arguments. Their proofs are provided in Sections 6.3–6.7.

The proof of Theorem 2 uses martingale approach of Stroock and Varadhan. Let us briefly recall the main ideas of this approach postponing the details till Section 6.7. According to [91] in order to show that $\mathcal{X}_\tau$ is a diffusion process with generator $\mathcal{L}$ it is enough to check that for a large set of observables B the process

$$\mathcal{M}_\tau = B(\mathcal{X}_\tau) - \int_0^\tau \mathcal{L}B(\mathcal{X}_\sigma)d\sigma$$

is a martingale. Thus one has to check that for sufficiently smooth functions $B_1, B_2 \ldots B_m$ and for all $s_1 \leq s_2 \cdots \leq s_m \leq \tau_1 \leq \tau_2$

$$\mathbb{E}\left(\left[\prod_{k=1}^m B_k(\mathcal{X}_{s_k})\right](\mathcal{M}_{\tau_2} - \mathcal{M}_{\tau_1})\right) = 0.$$

Therefore in order to show that a family of random processes $\{\mathcal{X}_\tau, M\}$ converges to $\mathcal{X}_\tau$ as $M \to \infty$ we need to show that

$$\mathbb{E}\left(\left[\prod_{k=1}^m B_k(\mathcal{X}_{s_k,M})\right](\mathcal{M}_{\tau_2,M} - \mathcal{M}_{\tau_1,M})\right) \to 0.$$

To derive this one usually divides the segment $[\tau_1, \tau_2]$ into small segments

$$\tau_1 = t_1 \leq t_2 \cdots \leq t_N = \tau_2$$

and uses Taylor development of B to estimate $B(\mathcal{X}_{t_{j+1},M}) - B(\mathcal{X}_{t_j,M})$. Thus one needs to control the moments $\mathbb{E}([\mathcal{X}_{t_{j+1},M} - \mathcal{X}_{t_j,M}]^p)$. Our first task is to control the moments of Q and V.

We need some notation. Let $\mathbf{n} = \kappa_M \sqrt{M}$, where for the proof of Theorem 2 we set $\kappa_M = M^{-\delta}$ with some $\delta > 0$, whereas for the proof of Theorem 1 we will need κ_M to be a small positive constant (independent of M).

The estimates of Propositions 3.2 and 3.3 require that $(Q, V) \in \Upsilon_{\delta_1}$, see (3.3), so we have to exclude the orbits leaving this region. Fix a $\delta_2 \in (\delta_1, \delta_0)$ and let $\ell = (\gamma, \rho)$ be a standard pair such that length$(\gamma) \geq M^{-100}$. We define, inductively, subsets
$$\emptyset = I_0 \subset I_1 \subset \ldots I_k \subset I_{k+1} \subset \cdots \subset \gamma,$$
which we will exclude from γ, as follows. Suppose that I_k is already defined so that

(i) $\mathcal{F}^{k\mathbf{n}}(\gamma \setminus I_k) = \bigcup_\alpha \gamma_{\alpha,k}$, where for each α we have length$(\gamma_{\alpha,k}) > M^{-100}$, and $\ell_{\alpha,k} = (\gamma_{\alpha,k}, \rho_{\alpha,k})$ is a standard pair (here $\rho_{\alpha,k}$ is the density of the measure $\mathcal{F}^{k\mathbf{n}}(\text{mes}_\ell)$ conditioned on $\gamma_{\alpha,k}$);

(ii) $\pi_1(\gamma \setminus I_k) \subset \Upsilon_{\delta_2}$, cf. (3.3).

Now, by Proposition 3.2, for each α
$$\mathcal{F}^{k\mathbf{n}} \gamma_{\alpha,k} = \Big(\bigcup_\beta \gamma_{\alpha,\beta,k+1}\Big) \bigcup \tilde{\gamma}_{\alpha,k+1}$$
where $\text{mes}_{\ell_{\alpha,k}}(\tilde{\gamma}_{\alpha,k+1}) < M^{-50}$ and length$(\gamma_{\alpha,\beta,k+1}) > M^{-100}$ for each β.

We now define
$$I_{k+1} = I_k \cup \Big(\cup_\alpha \mathcal{F}^{-(k+1)\mathbf{n}}(\tilde{\gamma}_{\alpha,k+1})\Big) \cup \Big(\cup^*_{\alpha,\beta} \mathcal{F}^{-(k+1)\mathbf{n}}(\gamma_{\alpha,\beta,k+1})\Big)$$
where $\cup^*_{\alpha,\beta}$ is taken over all pairs α, β such that $\pi_1(\gamma_{\alpha,\beta,k+1}) \notin \Upsilon_{\delta_2}$. For each $x \in \ell$ let $\mathbf{k}(x) = \min\{k \colon x \in I_k\}$ (we set $\mathbf{k}(x) = \infty$ if $x \notin I_k$ for any k).

For brevity, for each $x \in \Omega$ we denote the point $\mathcal{F}^n(x)$ by $x_n = (Q_n, V_n, q_n, v_n)$. Given a standard pair $\ell = (\gamma, \rho)$ as above, we define for every $x \in \gamma$

(6.1) $\quad \hat{Q}_n = \begin{cases} Q_n & \text{for } n < \mathbf{kn} \\ Q_{\mathbf{kn}} & \text{for } n \geq \mathbf{kn} \end{cases} \qquad \hat{V}_n = \begin{cases} V_n & \text{for } n < \mathbf{kn} \\ 0 & \text{for } n \geq \mathbf{kn} \end{cases}$

Recall that Theorem 2 claims a weak convergence of the stochastic processes $\mathcal{Q}(\tau)$ and $\mathcal{V}(\tau)$ on any finite time interval $0 < \tau < c$. From now on, we fix $c > 0$ and set $\bar{c} = c/\bar{L}$, where $\bar{L}$ is the mean free path (1.17).

Let $A \in \mathfrak{R}$ be a function satisfying two additional requirements:
(a) $\mu_{Q,V}(A) = 0$ for all Q, V;
(b) $\mu_{Q,V}(A^k)$ is a Lipschitz continuous function of Q and V for $k = 2, 3, 4$.

Given a standard pair $\ell = (\gamma, \rho)$ as above and $x \in \gamma$, we put $A(x_j) = A(\mathcal{F}^j x)$ and $S_n(x) = \sum_{j=0}^{n-1} A(x_j)$. We also set

$$\hat{A}(x_j) = \begin{cases} A(x_j) & \text{if } x \notin I_{[j/n]} \\ 0 & \text{otherwise} \end{cases} \quad \text{and} \quad \hat{S}_n = \sum_{j=0}^{n-1} \hat{A}(x_j)$$

cf. (6.1). In Section 6.4 we will prove

PROPOSITION 6.1. *The following bounds hold uniformly in $M^{1/2} \leq n \leq \bar{c} M^{2/3}$ and for all standard pairs $\ell = (\gamma, \rho)$ such that*

(6.2) $\quad \pi_1(\gamma) \subset \Upsilon^*_{\delta_2, a} := \{\text{dist}(Q, \partial \mathcal{D}) > \mathbf{r} + \delta_2, \ \|V\| < aM^{-2/3}\}$

and length$(\gamma) > M^{-100}$:

(a) $\mathbb{E}_\ell(\hat{S}_n) = \mathcal{O}(M^\delta n/\mathbf{n})$.

(b) $\mathbb{E}_\ell(\hat{S}_n^2) = \mathcal{O}(n)$.

(c) $\mathbb{E}_\ell(\hat{S}_n^4) = \mathcal{O}(n^2)$.

(d) *The last estimate can be specified as follows. Let*

(6.3) $\qquad \mathfrak{S}_A = \max_{(Q,V) \in \Upsilon_{\delta_2}} \max \{\mu_{Q,V}(A^2), D_{Q,V}(A)\}$

where

(6.4) $\qquad D_{Q,V}(A) = \sum_{j=-\infty}^{\infty} \mu_{Q,V}\left(A(A \circ \mathcal{F}^j_{Q,V})\right)$

Then

$$\mathbb{E}_\ell\left(\hat{S}_n^4\right) \leq 2\mathfrak{S}_A^2 n^2 + \mathcal{O}\left(n^{1.9}\right).$$

We remark that part (d) will only be used in Chapter 8, in the proof of Theorem 3.

Note that our functions $\hat{Q}_n$, $\hat{V}_n$, $\hat{A}_n$ and $\hat{S}_n$ are only defined on the selected standard pair ℓ and not on the entire phase space Ω yet. Given an auxiliary measure $m \in \mathfrak{M}$, we can use the corresponding partition of Ω into standard pairs $\{\ell = (\gamma, \rho)\}$, see Section 3.3, and define our functions on all the pairs $\{(\gamma, \rho)\}$ with length$(\gamma) > M^{-100}$, and then simply set these functions to zero on the shorter standard pairs. Now our functions are defined on Ω (but they depend on the measure $m \in \mathfrak{M}$ and the decomposition (3.4)).

Next, given $m \in \mathfrak{M}$ and $x \in \Omega$, we define continuous functions $\tilde{Q}(\tau)$, $\tilde{V}(\tau)$, and $\tilde{S}(\tau)$ on the interval $(0, \bar{c})$ by

(6.5) $\quad \tilde{Q}(\tau) = \hat{Q}_{\tau M^{2/3}}, \quad \tilde{V}(\tau) = M^{2/3} \hat{V}_{\tau M^{2/3}}, \quad \tilde{S}(\tau) = M^{-1/3} \hat{S}_{\tau M^{2/3}}$

(these formulas apply whenever $\tau M^{2/3} \in \mathbb{Z}$, and then we use linear interpolation in between). In a similar way, let t_n be the time of the nth collision, $\hat{t}_n = t_{\min\{n,\mathbf{kn}\}}$ the modified time, and then we define a continuous function

$$\tilde{t}(\tau) = M^{-1/3}\left[\hat{t}_{[\tau M^{2/3}]} - \bar{L}\min\{\tau M^{2/3}, \mathbf{kn}\}\right] \tag{6.6}$$

where $\bar{L}$ is the mean free path, cf. (1.17). We note that our normalization factors in (6.5)–(6.6) are chosen so that the resulting functions typically take values of order one, as we prove next.

Let us fix $a > 0$ and for each $M > 1$ choose an auxiliary measure $m \in \mathfrak{M}$ such that

$$m\left(\pi_1^{-1}\left(\Upsilon^*_{\delta_2,a}\right)\right) = 1, \tag{6.7}$$

see (6.2). Then each function $\tilde{S}(\tau), \tilde{Q}(\tau), \tilde{V}(\tau), \tilde{t}(\tau)$ induces a family of probability measures (parameterized by M) on the space of continuous functions $C[0,\bar{c}]$. We will investigate the tightness of these families. All our statements and subsequent estimates will be uniform over the choices of auxiliary measures $m \in \mathfrak{M}$ satisfying (6.7).

In Section 6.5 we will prove

PROPOSITION 6.2.
(a) *For every function $A \in \mathfrak{R}$ satisfying the assumptions of Proposition 6.1, the family of functions $\tilde{S}(\tau)$ is tight;*
(b) *the families $\tilde{Q}(\tau), \tilde{V}(\tau),$ and $\tilde{t}(\tau)$ are tight.*

COROLLARY 6.3. *For every sequence $M_k \to \infty$ there is a subsequence $M_{k_j} \to \infty$ along which the functions $\tilde{Q}(\tau)$ and $\tilde{V}(\tau)$ on the interval $0 < \tau < \bar{c}$ weakly converge to some stochastic processes $\hat{\mathbf{Q}}(\tau)$ and $\hat{\mathbf{V}}(\tau)$, respectively.*

Our next step is to use the tightness of $\hat{Q}$ and $\hat{V}$ to improve the estimates of Proposition 6.1(b). In Section 6.6 we will establish

PROPOSITION 6.4. *Let $\varkappa \ll \bar{c}$ be a small positive constant and $n = \varkappa M^{2/3}$. The following estimates hold uniformly for all standard pairs $\ell = (\gamma, \rho)$ such that $\pi_1(\gamma) \subset \Upsilon^*_{\delta_2,a}$ and $\text{length}(\gamma) > M^{-100}$, and all $(\bar{Q}, \bar{V}) \in \pi_1(\gamma)$:*

(a) $\mathbb{E}_\ell(\hat{V}_n - \bar{V}) = \mathcal{O}\left(\dfrac{1}{M^{1/3-\delta}\mathbf{n}}\right).$

(b) $\mathbb{E}_\ell\left((\hat{V}_n - \bar{V})(\hat{V}_n - \bar{V})^T\right) = \left(\bar{\sigma}^2_{\bar{Q}}(\mathcal{A}) + o_{\varkappa \to 0}(1)\right)\varkappa M^{-4/3}$

(c) $\mathbb{E}_\ell\left(\|\hat{V}_n - \bar{V}\|^4\right) = \mathcal{O}\left(\varkappa^2 M^{-8/3}\right)$

(d) $\mathbb{E}_\ell\left(\hat{Q}_n - \bar{Q}\right) = (1+o_{\varkappa\to 0}(1))\varkappa M^{2/3}\bar{L}\bar{V}+\mathcal{O}\left(\varkappa^{3/2}\right),$

(e) $\mathbb{E}_\ell\left(\|\hat{Q}_n - \bar{Q}\|^2\right) = \mathcal{O}\left(\varkappa^2\right).$

Let us now fix some $\delta_3 \in (\delta_2, \delta_0)$. Let $B(Q, V)$ be a C^3 smooth function of Q and V with a compact support whose projection on the Q space lies within the domain $\mathrm{dist}(Q, \partial \mathcal{D}) > \mathbf{r} + \delta_3$. Define a new function $\mathcal{L}B(Q, V)$ by

$$(\mathcal{L}B)(Q, V) = \bar{L}\langle V, \nabla_Q B\rangle + \frac{1}{2}\sum_{i,j=1}^{2}\left(\bar{\sigma}_Q^2(\mathcal{A})\right)_{ij}\partial^2_{V_i, V_j}B$$

where V_1 and V_2 denote the components of the vector V, and $\left(\bar{\sigma}_Q^2(\mathcal{A})\right)_{ij}$ stand for the components of the matrix $\bar{\sigma}_Q^2(\mathcal{A})$. As before, for each $M > 1$ we choose a measure $m \in \mathfrak{M}$ satisfying (6.7). In Section 6.7 we will prove

PROPOSITION 6.5. *Let* $(\hat{\mathbf{Q}}, \hat{\mathbf{V}})$ *be a stochastic process that is a limit point, as $M \to \infty$, of the family of functions $(\tilde{Q}(\tau), \tilde{V}(\tau))$ constructed above. Then the process*

$$\mathbf{M}(\tau) = B(\hat{\mathbf{Q}}(\tau), \hat{\mathbf{V}}(\tau)) - \int_0^\tau (\mathcal{L}B)(\hat{\mathbf{Q}}(s), \hat{\mathbf{V}}(s))\,ds$$

is a martingale.

Proposition 6.5 implies, in virtue of [**91**, Theorem 4.5.2], that any limit process $(\hat{\mathbf{Q}}, \hat{\mathbf{V}})$ satisfies

$$\begin{aligned} d\hat{\mathbf{Q}} &= \bar{L}\hat{\mathbf{V}}\,d\tau, & \hat{\mathbf{Q}}(0) &= Q_0 \\ d\hat{\mathbf{V}} &= \bar{\sigma}_{\hat{\mathbf{Q}}}(\mathcal{A})\,d\mathbf{w}(\tau), & \hat{\mathbf{V}}(0) &= 0 \end{aligned}$$

We will need an analogue of the above result for continuous time. For each $x \in \Omega$ consider two continuous functions on $(0, \bar{c}\bar{L})$ defined by

$$\tilde{Q}_*(\tau) = \begin{cases} Q(\tau M^{2/3}) & \text{for } \tau < \tilde{\tau} \\ Q(\tilde{\tau}) & \text{for } \tau \geq \tilde{\tau} \end{cases}$$

and

$$\tilde{V}_*(\tau) = \begin{cases} M^{2/3}V(\tau M^{2/3}) & \text{for } \tau < \tilde{\tau} \\ 0 & \text{for } \tau \geq \tilde{\tau} \end{cases}$$

where

(6.8) $\qquad \tilde{\tau} = M^{-2/3}\inf\{t > 0\colon (Q(t), V(t)) \notin \Upsilon_{\delta_2}\}$

In Section 6.7 we will derive

COROLLARY 6.6. *Suppose that $(\tilde{Q}(\tau), \tilde{V}(\tau))$ converges along some subsequence $M_k \to \infty$ to a process $(\hat{\mathbf{Q}}(\tau), \hat{\mathbf{V}}(\tau))$. Then $(\tilde{Q}_*(\tau), \tilde{V}_*(\tau))$ converges along the same subsequence $\{M_k\}$ to*

$$(\mathbf{Q}_*(\tau), \mathbf{V}_*(\tau)) = (\hat{\mathbf{Q}}(\tau/\bar{L}), \hat{\mathbf{V}}(\tau/\bar{L})).$$

Corollary 6.6 implies that $(\mathbf{Q}_*, \mathbf{V}_*)$ satisfies (2.18) up to the moment when $\mathrm{dist}(\mathbf{Q}_*(\tau), \partial \mathcal{D}) = \mathbf{r} + \delta_3$. We can now prove Proposition 3.5 (a). Observe that the difference between the limit process $\mathbf{Q}_*(\tau)$ above and the $\mathbf{Q}(\tau)$ involved in Theorem 2 is only due to the different stopping rules (6.8) and (2.12), respectively. In particular, $\mathbf{Q}_*$ can be stopped earlier than $\mathbf{Q}$ if for some $t \leq \bar{c}\bar{L}M^{2/3}$ we have $MV^2(t) \geq 1 - \delta_2$ but $\mathrm{dist}(Q(s), \partial \mathcal{D}) \geq \mathbf{r} + \delta_0$ for all $s \leq t$. By Proposition 6.2 (b) the probability of this event vanishes as $M \to \infty$. Thus any $(\mathbf{Q}, \mathbf{V})$ is obtained from the corresponding $(\mathbf{Q}_*, \mathbf{V}_*)$ by stopping the trajectory of the latter as soon as $\mathrm{dist}(\mathbf{Q}_*(\tau), \partial \mathcal{D}) = \mathbf{r} + \delta_0$. This fact concludes the proof of the main claim of Proposition 3.5 (a). Its part (b) will be proved in Section 6.9. □

6.2. Structure of the proofs. After having completed a formal description of all the intermediate steps in the proof of Proposition 3.5, let us give an informal overview of the underlying ideas.

Our argument derives from the martingale method of Stroock and Varadhan [91], which is based on the estimation of the first two moments of $V_n - V_0$. These are provided by Proposition 6.4, especially its part (b) saying that $\forall u \in \mathbb{R}^2$

$$(6.9) \qquad \mathbb{E}_\ell \left(\langle V_n - V_0, u \rangle^2 \right) \sim nM^{-2} \langle \bar{\sigma}_{Q_0}^2(\mathcal{A})u, u \rangle$$

for $n \sim \varkappa M^{2/3}$ (see our discussion in Section 2.2 for the motivation of this identity). Our proof of (6.9) proceeds in two steps: first we show that

$$(6.10) \qquad \mathbb{E}_\ell \left(\langle V_n - V_0, u \rangle^2 \right) \sim M^{-2} \sum_{j=0}^{n-1} \langle \bar{\sigma}_{Q_j}^2(\mathcal{A})u, u \rangle$$

(cf. Lemma 6.12), and then we approximate

$$(6.11) \qquad Q_j \sim Q_0$$

(cf. Proposition 6.2).

For any fixed j, the approximation (6.11) follows from

$$(6.12) \qquad \mathbb{E}_\ell \left(\|V_n - V_0\|^2 \right) \leq \mathrm{Const}\, n/M^2$$

by the Cauchy–Schwartz inequality. In order to get (6.11) for all j uniformly, we need to control the fourth moment, which can be derived from (6.10) with little difficulty. In turn, (6.12) itself follows from (6.10), hence the key step is to establish (6.10).

The proof of (6.10) is essentially based on the equidistribution (Proposition 3.3 and Corollary 3.4). For small n, Proposition 3.3 suffices. However, for $n \sim M^{2/3}$ the term $n\|\bar{V}\|$ becomes of order 1, so the estimates of Proposition 3.3 alone are too crude. In that case we first establish (6.10) for "short term", $n \sim M^{1/2-\delta}$ (Section 6.3), and then derive (6.10) for "long term", $n \sim M^{2/3}$, via Corollary 3.4 and the inductive estimate

$$\mathbb{E}_\ell\left(\|V_{(j-1)M^{1/2-\delta}}\|\right) \leq \text{Const } \frac{\sqrt{(j-1)M^{1/2-\delta}}}{M},$$

see details in Section 6.4.

Finally let us comment on the above inductive step. Denote $A^{(u)} = \langle \mathcal{A}, u \rangle$ for $u \in \mathbb{R}^2$ and consider the expansion

(6.13) $$\mathbb{E}_\ell\left(\left[\sum_{j=1}^n A^{(u)}(x_j)\right]^2\right) = \sum_{i,j=1}^n \mathbb{E}_\ell\left(A^{(u)}(x_i)A^{(u)}(x_j)\right)$$

Our early estimate (6.10) effectively states that the main contribution to (6.13) comes from nearly diagonal ($i \approx j$) terms. Thus to prove (6.10), it will suffice to bound the contribution of the off-diagonal terms in (6.13). There are two possible approaches to this task:

(I) Use Corollary 3.4 to estimate $\mathbb{E}_\ell\left[(A^{(u)}(x_i)A^{(u)}(x_j)\right]$. Since we expect the change of V to be of order $M^{-2/3}$, the best estimate we can get in this way is $\mathbb{E}_\ell\left[(A^{(u)}(x_i)A^{(u)}(x_j)\right] = \mathcal{O}(M^{-2/3}\ln M)$. Because there are $[M^{2/3}]^2$ terms in (6.13), this approach would provide an off-diagonal bound of $M^{2/3}\ln M$, which is way too crude – it is even larger than the main term $\mathcal{O}(M^{2/3})$.

(II) For a fixed i, we can try to get the inductive bound

$$\mathbb{E}_\ell\left(\sum_j A^{(u)}(x_j)A^{(u)}(x_i)\right) \leq \mathcal{O}\left(\sqrt{\# \text{ of terms}}\right) = \mathcal{O}(M^{1/3}).$$

This would give an off-diagonal bound of $\mathcal{O}(M^{2/3+1/3}) = \mathcal{O}(M)$, which is even worse...

Hence neither approach alone seems to handle the task, but they can be combined together to produce the necessary bound, in the framework of the so called "big-small block techniques". Namely, we divide the interval $[1, n]$ into "big" blocks of size $M^{1/2-\delta}$ separated by "small" blocks of size M^δ. The total contribution of the small blocks is negligible, and denoting by P'_j the contribution of the jth big block and setting $U'_k = \sum_{j=1}^{k} P'_j$ we can get

$$\mathbb{E}_\ell \left[(U'_{k+1})^2 - (U'_k)^2\right] = \mathbb{E}_\ell \left[(P'_{k+1})^2\right] + 2\,\mathbb{E}_\ell \left[U'_k P'_{k+1}\right].$$

The first term here can be handled by the method (I), while for the cross-product term we get, by Proposition 3.3,

$$\left|\mathbb{E}_\ell(U'_k P'_{k+1})\right| \leq \operatorname{Const} \mathbb{E}_\ell(|U'_k|)\,\mathbb{E}_\ell(P'_k),$$

and apply the method (II) to show that the first factor is of order $\sqrt{k M^{1/2+\delta}}$, while the second factor is of order 1 by the method (I). This approach yields the necessary bound on the off-diagonal terms in (6.13) and thus proves (6.10).

6.3. Short term moment estimates for V. Here we estimate the moments of the velocity V during time intervals of length $n = \mathcal{O}(\sqrt{M})$, which are much shorter than $\mathcal{O}(M^{2/3})$ required for Theorem 2. Our estimates will be used later in the proof of Proposition 6.1. The main result of this subsection is

PROPOSITION 6.7. *Let $\ell = (\gamma, \rho)$ be a standard pair such that $\pi_1(\gamma) \subset \Upsilon_{\delta_1}$ and $\operatorname{length}(\gamma) > M^{-100}$. Then for all $(\bar{Q}, \bar{V}) \in \pi_1(\gamma)$ and $M^{1/3} \leq n \leq \delta_\circ M^{1/2}$ we have*

(a) $\mathbb{E}_\ell(V_n - \bar{V}) = \mathcal{O}\left(M^{\delta-1}\right)$

(b) $\mathbb{E}_\ell(\|V_n - \bar{V}\|^2) = \mathcal{O}\left(n/M^2\right).$

Here $\delta_\circ \ll \delta_1$ is the constant of Proposition 3.3.

Proof. Let $\Delta V_j = V_{j+1} - V_j$. Then by (1.9)

$$(6.14) \qquad \Delta V_j = \frac{\mathcal{A} \circ \mathcal{F}^j}{M} + \mathcal{O}\left(\frac{1}{M^{3/2}}\right)$$

where $\mathcal{A}$ is as defined by (1.12). Hence

$$V_n - \bar{V} = \frac{1}{M} \sum_{i=0}^{n-1} \mathcal{A} \circ \mathcal{F}^i + \mathcal{O}\left(\frac{n}{M^{3/2}}\right)$$

and then

$$\|V_n - \bar{V}\|^2 \leq \frac{2}{M^2} \left\|\sum_{i=0}^{n-1} \mathcal{A} \circ \mathcal{F}^i\right\|^2 + \mathcal{O}\left(\frac{n^2}{M^3}\right).$$

Therefore Proposition 6.7 follows from the next result:

6. MOMENT ESTIMATES

PROPOSITION 6.8. *Let $A \in \mathfrak{R}$ be a function satisfying $\mu_{Q,V}(A) = 0$ for all Q, V, and ℓ and n be as in Proposition 6.7. Then*

(a) $\mathbb{E}_\ell(S_n) = \mathcal{O}\left(M^\delta\right)$

(b) $\mathbb{E}_\ell(S_n^2) = \mathcal{O}(n)$.

The proof uses the big small block techniques [3]. For each $k = 0, \ldots, [n/M^{1/3}]$ denote

$$R'_k = \sum_{j=kM^{1/3}+M^\delta}^{(k+1)M^{1/3}} A(x_j), \qquad R''_k = \sum_{j=kM^{1/3}}^{kM^{1/3}+M^\delta-1} A(x_j),$$

$$Z'_k = \sum_{j=0}^{k-1} R'_j, \qquad Z''_k = \sum_{j=0}^{k-1} R''_j.$$

Observe that $Z''_k \leq \|A\|_\infty n/M^{1/3-\delta}$. Next we prove two lemmas:

LEMMA 6.9. *For every k,*

(a) $\mathbb{E}_\ell(R'_k) = \mathcal{O}\left(M^{1/3+\delta}\left(\|\bar{V}\| + n/M\right)\right),$

(b) $\mathbb{E}_\ell\left([R'_k]^2\right) = \mathcal{O}\left(M^{1/3}\right).$

LEMMA 6.10. *Given A as above, there exists $D > 0$ such that*

(a) $\mathbb{E}_\ell(Z'_{k+1}) = \mathbb{E}_\ell(Z'_k) + \mathcal{O}\left(M^{1/3+\delta}\left(\|\bar{V}\| + n/M\right)\right),$

(b) $\mathbb{E}_\ell\left([Z'_{k+1}]^2\right) = \mathbb{E}_\ell\left([Z'_k]^2\right) + \mathcal{O}\left(M^{1/3}\right) + \mathcal{O}(\sqrt{k+1}\, M^{1/2+\delta}(\|\bar{V}\| + n/M))$ *and* $\mathbb{E}_\ell\left([Z'_k]^2\right) \leq DM^{1/3}(k+1)$.

Proof of Lemma 6.9. Applying Corollary 3.4 to $n_1 \leq n$ iterations of $\mathcal{F}$, setting $j = M^{\delta/4}$, and using the obvious bound

$$\|V_{n_1-j}\| \leq \|\bar{V}\| + \text{Const}\, n_1/M$$

we get

$$\mathbb{E}_\ell(A(x_{n_1})) = \mathcal{O}\left(\left(\|\bar{V}\| + n_1/M\right) M^{\delta/2}\right) + \mathcal{O}\left(M^{3\delta/4-1}\right)$$

Now (a) follows by summation over $kM^{1/3} + M^\delta \leq n_1 \leq (k+1)M^{1/3}$.

To prove (b) we write

$$(6.15) \qquad (R'_k)^2 = \sum_{i,j} A(x_i)A(x_j) = 2\sum_{i<j} A(x_i)A(x_j) + \mathcal{O}\left(M^{1/3}\right).$$

Thus it suffices to show that

$$(6.16) \qquad |\mathbb{E}_\ell\left(A(x_i)A(x_j)\right)| < \text{Const}\left(\theta_A^{j-i} + \left(\|\bar{V}\| + n/M\right) M^\delta\right)$$

for some $\theta_A < 1$. To prove (6.16) we apply Proposition 3.2 with $n = (i+j)/2$. Denoting $m = (j-i)/2$ we obtain

$$\mathbb{E}_\ell\left(A(x_i)A(x_j)\right) = \sum_\alpha c_\alpha \mathbb{E}_{\ell_\alpha}\left(A(x_{-m})A(x_m)\right).$$

If length$(\gamma_\alpha) > \exp(-m/K)$ where K is the constant of Proposition 3.2, choose $\bar{x}_\alpha \in \gamma_\alpha$. Due to (3.13) and the Hölder continuity of A, for any $x_\alpha \in \gamma_\alpha$ we have $|A(\mathcal{F}^{-m}x_\alpha) - A(\mathcal{F}^{-m}\bar{x}_\alpha)| = \mathcal{O}(\theta_A^m)$ for some constant $\theta_A < 1$, therefore

(6.17) $\quad \mathbb{E}_{\ell_\alpha}\left(A(x_{-m})A(x_m)\right) = A(\mathcal{F}^{-m}\bar{x}_\alpha)\,\mathbb{E}_{\ell_\alpha}\left(A(x_m)\right) + \mathcal{O}(\theta_A^m).$

By the argument used in the proof of Lemma 6.9 (a)

$$\mathbb{E}_{\ell_\alpha} A(x_m) = \mathcal{O}\left((\|\bar{V}\| + m/M)M^{\delta/2}\right),$$

hence

$$\mathbb{E}_{\ell_\alpha}\left(A(x_{-m})A(x_m)\right) = \mathcal{O}\left(\theta_A^m + \left(\|\bar{V}\| + m/M\right)M^\delta\right).$$

On the other hand, the contribution of α's which satisfy length$(\gamma_\alpha) \le \exp(-m/K)$ is exponentially small due to Proposition 3.2. Summation over α gives (6.16). Lastly, the summation over i,j and remembering that $n \le \delta_\diamond \sqrt{M}$ and $\|\bar{V}\| < 1/\sqrt{M}$ yields Lemma 6.9 (b). $\square$

Proof of Lemma 6.10. The part (a) follows directly from Lemma 6.9 (a). To prove part (b) we expand

(6.18) $\quad \mathbb{E}_\ell\big([Z'_{k+1}]^2\big) = \mathbb{E}_\ell\big([Z'_k]^2\big) + \mathbb{E}_\ell\big([R'_{k+1}]^2\big) + \mathbb{E}_\ell\big(Z'_k R'_{k+1}\big).$

The second term is $\mathcal{O}(M^{1/3})$ by Lemma 6.9 (b). We will show that the last term is much smaller, precisely

(6.19) $\quad \mathbb{E}_\ell\big(Z'_k R'_{k+1}\big) = \mathcal{O}\big(M^{\frac{1}{12}+\delta}\big)$

The argument used in the proof of (6.16) gives

$$\big|\mathbb{E}_\ell(Z'_k A_j)\big| \le \text{Const}\,\mathbb{E}_\ell|Z'_k|\left(\theta_A^{M^\delta} + \left(\|\bar{V}\| + n/M\right)M^\delta\right)$$

for $(k+1)M^{1/3} \le j \le (k+2)M^{1/3}$. Hence

$$\big|\mathbb{E}_\ell\big(Z'_k R'_{k+1}\big)\big| \le \text{Const}\,\mathbb{E}_\ell|Z'_k|\left(\theta_A^{M^\delta} + \left(\|\bar{V}\| + n/M\right)M^\delta\right)M^{1/3}$$

$$\le \text{Const}\,\sqrt{\mathbb{E}_\ell\big([Z'_k]^2\big)}\left(\|\bar{V}\| + n/M\right)M^{1/3+\delta}.$$

(where we used the Cauchy-Schwartz inequality). By induction

$$\mathbb{E}_\ell(Z'_k R'_{k+1}) \le C\sqrt{D}\Big(\sqrt{(k+1)M^{1/3}}\left(\|\bar{V}\| + n/M\right)\Big)M^{1/3+\delta}.$$

Since $k \leq \delta_\circ M^{1/6}$, the the right hand is $\mathcal{O}(M^{1/12+\delta})$. If D is sufficiently large, this implies both inequalities of part (b) for $k+1$ and thus completes the proof of Lemma 6.10. □

Proof of Proposition 6.8. To simplify our analysis we assume that $n = kM^{1/3}$ for some integer k, so that $S_n = Z'_k + Z''_k$. Similarly to the proof of Lemma 6.9 (a) we get

$$\mathbb{E}_\ell(R''_j) = \mathcal{O}\left(M^\delta(\|\bar{V}\| + n/M)\right)$$

for all $1 \leq j \leq k-1$ and

$$\mathbb{E}_\ell(R''_0) = \mathcal{O}\left(M^\delta\right).$$

(The difference between the first term and the others is due to the restriction $n > K|\ln \text{length}(\gamma)|$ in Proposition 3.2.) Combining the above estimates with Lemma 6.9 (a) we obtain part (a) of Proposition 6.8. To prove part (b) we estimate

$$\mathbb{E}_\ell\left(S_n^2\right) \leq 2\,\mathbb{E}_\ell\left([Z'_k]^2\right) + 2\,\mathbb{E}_\ell\left([Z''_k]^2\right) = \mathcal{O}(n + k^2 M^{2\delta}).$$

This completes the proof of Proposition 6.8 and hence that of 6.7.

6.4. Moment estimates–*a priori* bounds. Here we prove Proposition 6.1.

First we get a useful bound on multiple correlations. Let $A_1, \ldots, A_p$ and $B_1, \ldots, B_q$ be some functions from our class $\mathfrak{R}$ and c_1, c_2 some constants. Consider the functions

$$A(x) = \sum_* A_1(x_{i_1}) \cdots A_p(x_{i_p}) - c_1, \quad B(x) = \sum_{**} B_1(x_{j_1}) \cdots B_q(x_{j_q}) - c_2$$

where the summations $\sum_*$ and $\sum_{**}$ are performed over two different sets of indices (time moments). Let m_* be the *maximal* index in the first set (denoted by $*$) and m_{**} the *minimal* index in the second set. We suppose that $m_* \leq m - M^\delta < m \leq m_{**}$ for some m, i.e. there is a "time gap" of length $\geq M^\delta$ between m_* and m_{**}.

Now, let $\ell = (\gamma, \rho)$ be a standard pair such that $\text{length}(\gamma) > M^{-100}$. For any function C, we can decompose the expectation

(6.20) $$\mathbb{E}_\ell(C \circ \mathcal{F}^{m-\frac{1}{2}M^\delta}) = \sum_\alpha \mathbb{E}_{\ell_\alpha}(C)$$

where ℓ_α denote the components of the image of ℓ under $\mathcal{F}^{m-\frac{1}{2}M^\delta}$.

LEMMA 6.11 (Multiple correlations). *We have*

$$\left|\mathbb{E}_\ell(A(x)B(x))\right| \leq \mathbb{E}_\ell|A(x)| \max_\alpha \left|\mathbb{E}_{\ell_\alpha}\left(B(x_{-m+\frac{1}{2}M^\delta})\right)\right| + \mathcal{O}\left(M^{-50}\right)$$

where the maximum is taken over α's in (6.20) with $\text{length}(\gamma_\alpha) > M^{-100}$. We note that the remainder term $\mathcal{O}(M^{-50})$ here depends on the choice of the functions A_i, B_j and the constants c_1, c_2.

The proof of Lemma 6.11 is similar to that of (6.16), in which the factorization (6.17) plays a key role, we omit details. $\square$

We now turn to the proof of Proposition 6.1. Using big small blocks again, we put for all $k = 0, \ldots, [\bar{c}M^{2/3}/\mathbf{n}]$

$$P'_k = \sum_{j=k\mathbf{n}+M^\delta}^{j=(k+1)\mathbf{n}} \hat{A}(x_j), \qquad P''_k = \sum_{j=k\mathbf{n}}^{k\mathbf{n}+M^\delta} \hat{A}(x_j),$$

and then

$$U'_k = \sum_{j=0}^{k-1} P'_j, \qquad U''_k = \sum_{j=0}^{k-1} P''_j.$$

Note that $U''_k = \mathcal{O}\left((k+1)M^\delta\right)$.

LEMMA 6.12. *Under the assumptions of Proposition 6.1 and uniformly in k*

(a) $\mathbb{E}_\ell(P'_k) = \mathcal{O}\left(M^\delta\right)$.

(b) $\mathbb{E}_\ell\left([P'_k]^2\right) = \left(\mathbb{E}_\ell(\hat{D}_{Q_{k\mathbf{n}}, V_{k\mathbf{n}}}) + g\right)\mathbf{n}$

where

$$\hat{D}_{Q_{k\mathbf{n}}, V_{k\mathbf{n}}}(x) = \begin{cases} D_{Q_{k\mathbf{n}}, V_{k\mathbf{n}}}(A) & \text{if } x \notin I_k \\ 0 & \text{otherwise} \end{cases}$$

see (6.4), and $g \to 0$ as $M \to \infty$ and $\kappa_M \to 0$, see a remark below.

(c) $\mathbb{E}_\ell\left([P'_k]^4\right) = \mathcal{O}\left(\mathbf{n}^2\right)$.

(d) *In the notation of (6.3), we have*

$$\mathbb{E}_\ell\left([P'_k]^4\right) \leq 2\mathfrak{S}_A^2 \mathbf{n}^2 + \mathcal{O}\left(\mathbf{n}^{1.9}\right).$$

Remark. In the proof of Theorem 2 we set $\kappa_M = M^{-\delta}$, hence $\kappa_M \to 0$ follows from $M \to \infty$, and so we can replace g in (b) by $o(1)$. In the proof of Theorem 1, however, κ_M will be a small constant (independent of M), hence the condition $\kappa_M \to 0$ will be necessary.

Proof. We first note that $\hat{A}$ is different from 0 only on standard pairs where (3.10) holds, see the construction of $\hat{A}$ in Section 6.1. Thus, Proposition 3.3 applies to each standard pair $\gamma_{\alpha,k}$ where $\hat{A} \neq 0$. Therefore it is enough to verify Lemma 6.12 for $k = 0$ (but we need to establish it for all $\ell = (\gamma, \rho)$ such that $\pi_1(\gamma) \subset \Upsilon_{\delta_2}$).

Part (a) follows from Proposition 6.8 (a).

6. MOMENT ESTIMATES

The proof of (b) is based on the following claim: for each $\varepsilon > 0$ there exists $K(\varepsilon)$ (it is enough to set $K(\varepsilon) = \text{Const} \,|\ln \varepsilon|$) such that

$$(6.21) \qquad \mathbb{E}_\ell \left(\sum_{|i-j|>K(\varepsilon)} \hat{A}(x_i)\hat{A}(x_j) \right) < \varepsilon \mathbf{n}$$

(here, of course, $M^\delta \leq i, j \leq \mathbf{n}$). To prove (6.21), we apply the big small block decomposition, as in Lemmas 6.9 and 6.10, to $[P_k']^2$ (with big blocks of length $M^{1/3}$ and small blocks of length M^δ), then we use Eqs. (6.18)–(6.19), the induction on k, and finally the estimate (6.16) applied to each big block will yield (6.21).

Therefore, to get the asymptotics of $\mathbb{E}_\ell([P_0']^2)$ we need to get the asymptotics of

$$\mathbb{E}_\ell \left(\sum_{i=M^\delta}^{\mathbf{n}} \hat{A}(x_i)\hat{A}(x_{i+m}) \right)$$

for each fixed m. Applying Proposition 3.2 to $j = i - M^\delta$ iterations of $\mathcal{F}$ we get

$$\mathbb{E}_\ell \big(\hat{A}(x_i)\hat{A}(x_{i+m}) \big) = \sum_\alpha c_\alpha \mathbb{E}_{\ell_\alpha} \big(\hat{A}(x_{M^\delta})\hat{A}(x_{M^\delta+m}) \big).$$

where $\ell_\alpha = (\gamma_\alpha, \rho_\alpha)$ denote the components of the image of ℓ at time j. Proposition 3.3 applies to each γ_α where $\hat{A}(x_{M^\delta}) \neq 0$, hence for each α such that length$(\gamma_\alpha) > \exp(-M^\delta/K)$ we have

$$\mathbb{E}_{\ell_\alpha} \big(\hat{A}(x_i)\hat{A}(x_{i+m}) \big) = \mu_{\bar{Q},\bar{V}} \big(\hat{A}(x_0)\hat{A}(x_m) \big) + \mathcal{O}\left(\|\bar{V}\|M^\delta + M^{2\delta-1} \right)$$

where $(\bar{Q},\bar{V}) \in \pi_1(\gamma_\alpha)$ is an arbitrary point. As before, the contribution of small γ_α is well within the error bounds of our claim (b), hence summing over α and using the fact that the oscillations of Q and V over γ_α are of order $1/M$ we obtain

$$\mathbb{E}_\ell \left(\hat{A}(x_i)\hat{A}(x_{i+m}) \right) = \mathbb{E}_\ell \left(\mu_{Q_j,V_j}(\hat{A}(x_0)\hat{A}(x_m)) \right)$$
$$+ \mathcal{O}\left(\mathbb{E}_\ell(\|\hat{V}_j\|)M^\delta + M^{2\delta-1} \right).$$

(recall that $j = i - M^\delta$). By Proposition 6.7

$$\mathbb{E}_\ell(\|\hat{V}_j\|) = \mathcal{O}(\|\bar{V}\| + \sqrt{i}/M) = \mathcal{O}(\|\bar{V}\| + M^{-3/4})$$

whereas by Lemma 4.17

$$\mathbb{E}_\ell \left(\mu_{Q_j,V_j}(\hat{A}(x_0)\hat{A}(x_m)) \right) = \mathbb{E}_\ell \left(\mu_{\bar{Q},\bar{V}}(\hat{A}(\hat{A} \circ \mathcal{F}_{\bar{Q},\bar{V}}^m)) \right)$$
$$+ \mathcal{O}\big(\mathbb{E}_\ell \|\hat{Q}_j - \bar{Q}\| \big)$$
$$+ \mathcal{O}\big(\mathbb{E}_\ell \|\hat{V}_j - \bar{V}\| \big) + \mathcal{O}\big(\|\bar{V}\| + M^{-1} \big)$$

Proposition 6.7 (b) gives
$$\mathbb{E}_\ell\big(\|\hat{V}_j - \bar{V}\|\big) \leq \text{Const } \sqrt{j}/M \leq \text{Const } M^{-3/4}$$
(we note that $\mathbf{n} = \kappa_M M^{1/2} < \delta_\diamond M^{1/2}$, hence Proposition 6.7 indeed applies in our context). Also, since $M\|V_j\|^2 < 1 - \delta_1$, then $\|v_j\| \geq \text{Const} > 0$, hence intercollision times are uniformly bounded above for all $j \leq \mathbf{kn}$. Therefore,
$$\mathbb{E}_\ell\big(\|\hat{Q}_j - \bar{Q}\|\big) \leq \text{Const} \sum_{p=0}^{j-1} \mathbb{E}_\ell\big(\|\hat{V}_p\|\big)$$
$$\leq \text{Const } j\,\|\bar{V}\| + \sum_{p=0}^{j-1} \mathbb{E}_\ell\big(\|\hat{V}_p - \bar{V}\|\big)$$
$$\leq \text{Const}\big(j\,\|\bar{V}\| + j\,M^{-3/4}\big).$$

Note that
$$j\,\|\bar{V}\| \leq \text{Const } \mathbf{n}/\sqrt{M} = \text{Const } \kappa_M, \qquad j\,M^{-3/4} \leq \kappa_M M^{-1/4}.$$
This gives
$$\mathbb{E}_\ell\big(\hat{A}(x_i)\hat{A}(x_{i+m})\big) = \mu_{\bar{Q},\bar{V}}\big(\hat{A}\,(\hat{A} \circ \mathcal{F}^m_{\bar{Q},\bar{V}})\big) + \mathcal{O}(\kappa_M).$$
We note that all our constants and the $\mathcal{O}(\cdot)$ terms depend, implicitly, on m which takes values between 0 and K_ε. Summing over i, m we get
$$\mathbb{E}_\ell\big([P'_0]^2\big) = \mathbf{n}\bigg(\sum_{|m|<K(\varepsilon)} \mu_{\bar{Q},\bar{V}}\big(\hat{A}\,(\hat{A} \circ \mathcal{F}^m_{\bar{Q},\bar{V}})\big) + \mathcal{R} + \mathcal{O}(\kappa_M)\bigg),$$
where $|\mathcal{R}| \leq \varepsilon$ and the $\mathcal{O}(\cdot)$ term implicitly depends on ε. Now for any $\varepsilon > 0$ we can choose a small enough κ_M so that the $|\mathcal{O}(\kappa_M)| < \varepsilon$. This concludes the proof of part (b).

We proceed to the proof of part (d). We write
$$\mathbb{E}_\ell\big([P'_0]^4\big) = \sum_{i_1,i_2,i_3,i_4} \mathbb{E}_\ell\big(\hat{A}(x_{i_1})\hat{A}(x_{i_2})\hat{A}(x_{i_3})\hat{A}(x_{i_4})\big).$$
For convenience, we order the indices in each term so that
$$(6.22) \qquad i_1 \leq i_2 \leq i_3 \leq i_4$$
There are eight cases depending on the choice of "$<$" or "$=$" in (6.22), but we will be able to handle several cases together. First we separate the terms in which $i_1 \leq i_2 < i_3 < i_4$ and get
$$(6.23) \qquad \mathbb{E}_\ell\big([P'_0]^4\big) = \sum_{m=M^\delta+1}^{\mathbf{n}} \mathbb{E}_\ell\bigg(\hat{S}^2_m\,\hat{A}(x_m) \sum_{j=m+1}^{\mathbf{n}} \hat{A}(x_j)\bigg) + \mathcal{R}$$

where we denote, for convenience, $j = i_4$, $m = i_3$ and $\hat{S}_m = \sum_{i=M^\delta}^{m-1} \hat{A}(x_i)$, while $\mathcal{R}$ correspond to all the remaining terms. Denote

$$\hat{S}^{(a)} = \hat{S}_{m-M^\delta}, \qquad \hat{S}^{(b)} = \hat{S}_m - \hat{S}_{m-M^\delta},$$
$$\hat{S}^{(c)} = \hat{S}_{m+M^\delta} - \hat{S}_m, \qquad \hat{S}^{(d)} = \hat{S}_\mathbf{n} - \hat{S}_{m+M^\delta},$$

then we have

$$\mathbb{E}_\ell \left(\hat{S}_m^2 \hat{A}(x_m) \sum_{j>m} \hat{A}(x_j) \right) = \mathbb{E}_\ell \left([\hat{S}^{(a)}]^2 \hat{A}(x_m) \sum_{j>m} \hat{A}(x_j) \right)$$

$$+ \mathbb{E}_\ell \left([\hat{S}^{(b)}]^2 \hat{A}(x_m) \sum_{j>m} \hat{A}(x_j) \right) + 2\mathbb{E}_\ell \left(\hat{S}^{(a)} \hat{S}^{(b)} \hat{A}(x_m) \sum_{j>m} \hat{A}(x_j) \right)$$

$$= I + I\!I + I\!I\!I.$$

We can assume here that $m > M^{3\delta}$, because terms with $m < M^{3\delta}$ make a total contribution of order $\mathbf{n} M^{9\delta}$. It is clear that part (b) of Lemma 6.12 can be applied to any number of iterations $M^{3\delta} < m \leq \mathbf{n}$, hence $\mathbb{E}_\ell \left([\hat{S}^{(a)}]^2 \right) < (\mathfrak{S}_A + o(1)) m$. Therefore by Lemma 6.11

$$|I| < m(\mathfrak{S}_A + o(1)) \max_\alpha \left| \mathbb{E}_{\ell_\alpha} \left(A(x_{M^\delta/2}) \sum_{j=1}^{\mathbf{n}-m} A(x_{j+M^\delta/2}) \right) \right| + \mathcal{O}\left(M^{-50} \right),$$

where ℓ_α denote the components of the image of ℓ at time $m - M^\delta/2$. Similar to the proof of Lemma 6.12 (b), for each α we have
(6.24)
$$\left| \mathbb{E}_{\ell_\alpha} \left(\hat{A}(x_{M^\delta/2}) \sum_{j=1}^{\mathbf{n}-m} \hat{A}(x_{j+M^\delta/2}) \right) \right| \leq \sum_{j=1}^\infty \mu_{\bar{Q},\bar{V}} \left(A \left(A \circ \mathcal{F}^j_{\bar{Q},\bar{V}} \right) \right) + o(1).$$

We observe that

$$\sum_{j=1}^\infty \mu_{\bar{Q},\bar{V}} \left(A \left(A \circ \mathcal{F}^j_{\bar{Q},\bar{V}} \right) \right) = \frac{D_{\bar{Q},\bar{V}}(A) - \mu_{\bar{Q},\bar{V}}(A^2)}{2} \leq \mathfrak{S}_A,$$

hence $I = \mathcal{O}\left(\mathfrak{S}_A^2 m \right)$. Next,

$$I\!I\!I = \mathbb{E}_\ell \left(\hat{S}^{(a)} \hat{S}^{(b)} \hat{A}(x_m) [\hat{S}^{(c)} + \hat{S}^{(d)}] \right)$$
$$= \mathbb{E}_\ell \left(\hat{S}^{(a)} \hat{S}^{(b)} \hat{A}(x_m) \hat{S}^{(c)} \right) + \mathbb{E}_\ell \left(\hat{S}^{(a)} \hat{S}^{(b)} \hat{A}(x_m) \hat{S}^{(d)} \right)$$
$$= I\!I\!I_c + I\!I\!I_d.$$

Now

$$|I\!I\!I_c| \leq \text{Const } M^{2\delta} \mathbb{E}_\ell \left(|S^{(a)}| \right) \leq \text{Const } M^{2\delta} \sqrt{m}$$

where the last inequality is based on Lemma 6.12 (b) and Cauchy-Schwartz. On the other hand, due to Lemma 6.11 and Proposition 6.8 (a)

$$|I\!I\!I_d| \leq \text{Const } M^\delta \mathbb{E}_\ell\left(\left|\hat{S}^{(a)}\hat{S}^{(b)}\hat{A}(x_m)\right|\right)$$
$$\leq \text{Const } M^{2\delta} \mathbb{E}_\ell(|\hat{S}^{(a)}|)$$
$$\leq \text{Const } M^{2\delta}\sqrt{m}$$

(here again the last inequality follows from Lemma 6.12 (b)). Thus $|I\!I\!I| \leq \text{Const } M^{2\delta}\sqrt{m}$. Similar estimates show that $|I\!I| \leq \text{Const } M^{3\delta}$. Combining these results we get

$$\left|\mathbb{E}_\ell\left(\hat{S}_m^2 \hat{A}(x_m) \sum_{j>m} \hat{A}(x_j)\right)\right| \leq 2\mathfrak{S}_A^2 m + \mathcal{O}\left(M^{2\delta}\sqrt{m} + M^{3\delta}\right).$$

Summation over m gives

$$\left|\mathbb{E}_\ell\left(\sum_m \hat{S}_m^2 \hat{A}(x_m) \sum_{j>m} \hat{A}(x_j)\right)\right| \leq 2\mathfrak{S}_A^2 \mathbf{n}^2 + \mathcal{O}\left(\mathbf{n}^{3/2+6\delta}\right).$$

It remains to estimate the term $\mathcal{R}$ in (6.23), which corresponds to the cases where $i_2 = i_3$ or $i_3 = i_4$. The cases where $i_1 \leq i_2 < i_3 = i_4$ can be treated in the same way as above, except (6.24) now takes form

$$\mathbb{E}_{\ell_\alpha}\left(\hat{A}^2(x_{M^\delta/2})\right) = \mu_{\bar{Q},\bar{V}}(A^2) + o(1) \leq \mathfrak{S}_A$$

and the term $I\!I\!I_d$ is missing altogether.

The case $i_1 = i_2 = i_3 < i_4$ and that of $i_1 < i_2 = i_3 = i_4$ can be handled as follows:

$$\left|\mathbb{E}_\ell\left(\sum_{i \neq k} \hat{A}(x_k)^3 \hat{A}(x_i)\right)\right| \leq \text{Const} \sum_k \mathbb{E}\left(|\hat{S}_\mathbf{n}| + 1\right) \leq \text{Const } \mathbf{n}^{3/2}.$$

Consider the case $i_1 < i_2 = i_3 < i_4$. Using the same notation as in the analysis of the first term in (6.23) we get

$$\sum_m \mathbb{E}_\ell\left[\left(\hat{S}^{(a)} + \hat{S}^{(b)}\right)\hat{A}^2(x_m)\left(\hat{S}^{(c)} + \hat{S}^{(d)}\right)\right] = \sum_m \left[I^{ac} + I^{ad} + I^{bc} + I^{bd}\right]$$

where we denoted $I^{\alpha\beta} = \mathbb{E}_\ell\left(\hat{S}^{(\alpha)}\hat{S}^{(\beta)}\hat{A}^2(x_m)\right)$. The estimation of each term here is similar to the ones discussed above, and we obtain

$$I^{ad} = \mathcal{O}\left(\sqrt{\mathbb{E}_\ell(\hat{S}^{(a)})^2}M^\delta + M^{-50}\right) = \mathcal{O}\left(\sqrt{m}\,M^\delta\right),$$
$$I^{ac} = \mathcal{O}\left(\sqrt{m}\,M^\delta\right),$$
$$I^{bc} = \mathcal{O}\left(M^{2\delta}\right),$$

6. MOMENT ESTIMATES

$$I^{bd} = \mathcal{O}\left(M^{2\delta}\right).$$

Hence
$$\sum_{i_1 < m < i_4} \mathbb{E}_\ell\left(\hat{A}(x_{i_1})\hat{A}^2(x_m)\hat{A}(x_{i_4})\right) = \mathcal{O}\left(\mathbf{n}^{3/2}\right).$$

The only remaining case $i_1 = i_2 = i_3 = i_4$ is simple:
$$\sum_j \mathbb{E}_\ell(\hat{A}^4(x_j)) \leq \text{Const } \mathbf{n}.$$

This proves part (d). Obviously, part (c) follows from (d), which completes the proof of Lemma 6.12. □

We now return to the proof of Proposition 6.1. Denote

(6.25) $$\mathbb{E}_{\max}(\dots) = \max_\ell |\mathbb{E}_\ell(\dots)|$$

where the maximum is taken over all standard pairs $\ell = (\gamma, \rho)$ with length$(\gamma) > M^{-100}$.

We are now going to prove by induction that

(6.26) $$\mathbb{E}_{\max}(U'_k) \leq G_1 k M^\delta,$$

(6.27) $$\mathbb{E}_{\max}([U'_k]^2) \leq G_2 k \mathbf{n},$$

(6.28) $$\mathbb{E}_{\max}([U'_k]^4) \leq G_4 k^2 \mathbf{n}^2$$

provided the constants G_1, G_2, G_4 are sufficiently large. Let us rewrite the estimates of Lemma 6.12 in a simplified way:

(6.29) $$\mathbb{E}_\ell(P'_k) \leq C_1 M^\delta, \quad \mathbb{E}_\ell\left([P'_k]^2\right) \leq C_2 \mathbf{n}, \quad \mathbb{E}_\ell\left([P'_k]^4\right) \leq C_4 \mathbf{n}^2.$$

Now, by the inductive assumption (6.26) we have
$$\left|\mathbb{E}_\ell\left(U'_{k+1}\right)\right| \leq G_1 k M^\delta + C_1 M^\delta,$$
hence (6.26) holds for $k+1$ provided that $G_1 > C_1$.

Next, by Lemma 6.11, (6.29), the inductive assumption (6.27), and the Cauchy-Schwartz inequality we have

$$\mathbb{E}_{\max}\left([U'_{k+1}]^2\right) \leq \mathbb{E}_{\max}\left([U'_k]^2\right) + 2\mathbb{E}_{\max}\left(U'_k P'_k\right) + \mathbb{E}_{\max}\left([P'_k]^2\right)$$

(6.30) $$\leq G_2 k \mathbf{n} + 2 C_1 M^\delta \sqrt{G_2 k \mathbf{n}} + C_2 \mathbf{n}.$$

Since $k\mathbf{n} < \bar{c} M^{2/3}$, the second term here is $\mathcal{O}(M^{-1/6+2\delta}\mathbf{n})$, hence (6.27) holds provided that $G_2 > C_2$.

Lastly, by the inductive assumption (6.28) we have

$$\mathbb{E}_{\max}\left([U'_{k+1}]^4\right) \leq \mathbb{E}_{\max}\left([U'_k]^4\right) + 4\mathbb{E}_{\max}\left([U'_k]^3 P'_k\right) + 6\mathbb{E}_{\max}\left([U'_k]^2[P'_k]^2\right)$$
$$+ 4\mathbb{E}_{\max}\left(U'_k[P'_k]^3\right) + \mathbb{E}_{\max}\left([P'_k]^4\right)$$
$$\leq G_4 k^2 \mathbf{n}^2 + I + II + III + IV.$$

Using Lemma 6.11, (6.29), and the Hölder inequality we get

$$I \leq 4C_1 G_4^{3/4}[k\mathbf{n}]^{3/2}M^\delta = 4C_1 G_4^{3/4} k\mathbf{n}^2 \sqrt{\frac{k}{\mathbf{n}}} M^\delta,$$
$$I\!I \leq 6(G_2 k\mathbf{n})(C_2\mathbf{n}) = 6(G_2 C_2)k\mathbf{n}^2,$$
$$I\!I\!I \leq 4\sqrt{G_2 k\mathbf{n}}\,(C_4^{3/4}\mathbf{n}^{3/2}) = \sqrt{G_2}\,C_4^{3/4}\sqrt{k}\,\mathbf{n}^2,$$
$$I\!V \leq C_4 \mathbf{n}^2.$$

Hence, (6.28) holds provided that $G_4 > 6C_2 G_2$. This completes the proof of (6.26)–(6.28) establishing the parts (a)–(c) of Proposition 6.1 (the contribution from U_k'' is well within our error bounds). The proof of (d) is similar to (c), but we have to use Lemma 6.12 (d) in place of Lemma 6.12 (c). Proposition 6.1 is proved. □

Let us also note, for future reference, that by (6.30) the main difference between $\mathbb{E}_\ell\left([U'_{k+1}]^2\right)$ and $\mathbb{E}_\ell\left([U'_k]^2\right)$ comes from the $[P'_k]^2$ term. Hence we have

$$(6.31) \qquad \mathbb{E}_\ell\left(\hat{S}^2_{\bar{c}M^{2/3}}\right) = \sum_{k \leq \bar{c}M^{2/3}/\mathbf{n}} \mathbb{E}_\ell\left([P'_k]^2\right) + \mathcal{O}\left(M^{1/3+2\delta}\right).$$

6.5. Tightness. We precede the proof of Proposition 6.2 with a few general remarks.

To establish the tightness of a family of probability measures $\{P_M\}$ on the space of continuous functions $C[0, \bar{c}]$ we need to show that for any $\varepsilon > 0$ there exists a compact subset $K_\varepsilon \subset C[0, \bar{c}]$ such that $P_M(K_\varepsilon) > 1 - \varepsilon$ for all M. The compactness of K_ε means that the functions $\{F \in K_\varepsilon\}$ are uniformly bounded at $\tau = 0$ and equicontinuous on $[0, \bar{c}]$. All our families of functions in Proposition 6.2 are obviously uniformly bounded at $\tau = 0$, hence we only need to worry about the equicontinuity. For any $M_0 > 0$ all our functions $\tilde{S}(\tau), \tilde{Q}(\tau), \tilde{V}(\tau)$, and $\tilde{t}(\tau)$ corresponding to $M < M_0$ have uniformly bounded derivatives (with a bound depending on M_0), hence they trivially make a compact set. Thus, to prove the tightness for these functions, it is enough to construct a compact set K_ε such that $P_M(K_\varepsilon) > 1 - \varepsilon$ for all $M > M_\varepsilon$ with some $M_\varepsilon > 1$, hence in our proofs we can (and will) assume that M is large enough.

Lastly, recall that each $m \in \mathfrak{M}$ satisfies (3.4)–(3.6). Since now we can assume that $M_\varepsilon^{-50} < \varepsilon/2$, it will be enough to prove all necessary measure estimates for measures mes_ℓ on individual standard pairs $\ell = (\gamma, \rho)$ with $\text{length}(\gamma) > M^{-100}$ (but our estimates must be uniform over all such standard pairs).

First we prove the part (a) of Proposition 6.2. Let $\mathcal{C}_N$ be the space of continuous functions $S(\tau)$ on $[0, \bar{c}]$ such that

$$\left| S\left(\frac{k+1}{2^m}\right) - S\left(\frac{k}{2^m}\right) \right| \leq 2^{-\frac{m}{8}} \tag{6.32}$$

for all $m \geq N$ and $k < 2^m \bar{c}$. Observe that functions in $\mathcal{C}_N$ are equicontinuous since they are uniformly Hölder on a dense set (of binary rationals) and by continuity they are globally Hölder continuous. We claim that for each $\mathcal{E} > 0$ there exists N such that for all $\ell = (\gamma, \rho)$ with length$(\gamma) > M^{-100}$

$$\text{mes}_\ell(\tilde{S} \in \mathcal{C}_N) > 1 - \mathcal{E}$$

uniformly in M, where $\tilde{S}$ is defined by (6.5). Note that

$$\left| \tilde{S}\left(\frac{k+1}{2^m}\right) - \tilde{S}\left(\frac{k}{2^m}\right) \right| \leq \frac{\|A\|_\infty M^{1/3}}{2^m} \tag{6.33}$$

so (6.32) holds for all m such that $2^{-m} < \text{Const}\, M^{-8/21}$. Assume now that $2^{-m} \geq \text{Const}\, M^{-8/21}$. Equivalently, we need to estimate $|\hat{S}_{n_2} - \hat{S}_{n_1}|$ for $|n_2 - n_1| \geq \text{Const}\, M^{2/7}$.

LEMMA 6.13. *For all n_1, n_2 such that $|n_2 - n_1| > \text{Const}\, M^{2/7}$ and for all $\ell = (\gamma, \rho)$ with* length$(\gamma) > M^{-100}$

$$\mathbb{E}_\ell\left([\hat{S}_{n_2} - \hat{S}_{n_1}]^4 \right) \leq \text{Const}\, (n_2 - n_1)^2.$$

Proof. For $n_2 - n_1 \geq cM^{1/2}$, our estimate follows from Lemma 6.12 (c) and the argument used in the proof of Proposition 6.1 (c). For smaller $n_2 - n_1$, the proof is similar to that of Lemma 6.12 (c). $\square$

Lemma 6.13 implies that for fixed k, m

$$\Delta_m := \text{mes}_\ell\left(\left| \tilde{S}\left(\frac{k+1}{2^m}\right) - \tilde{S}\left(\frac{k}{2^m}\right) \right| > 2^{-m/8} \right)$$

$$= \text{mes}_\ell\left(\left| \tilde{S}\left(\frac{k+1}{2^m}\right) - \tilde{S}\left(\frac{k}{2^m}\right) \right|^4 > 2^{-m/2} \right)$$

$$\leq 2^{m/2} \mathbb{E}_\ell\left(\left| \tilde{S}\left(\frac{k+1}{2^m}\right) - \tilde{S}\left(\frac{k}{2^m}\right) \right|^4 \right)$$

$$\leq \text{Const}\, \frac{2^{m/2}}{2^{2m}} = \text{Const}\, 2^{-3m/2}.$$

Summation over k and m completes the proof of part (a) of Proposition 6.2.

We now prove part (b). The tightness of $\tilde{V}(\tau)$ follows from (1.9), (6.14) and Proposition 6.2 (a) applied to the function $\mathcal{A}$ (the contribution of the correction term $\mathcal{O}\left(M^{-3/2}\right)$ in (6.14) is well within our error bounds).

The equicontinuity of $\tilde{Q}(\tau)$ follows from a simple estimate:

$$\|\tilde{Q}(\tau_2) - \tilde{Q}(\tau_1)\| \leq (\tau_2 - \tau_1) \max_\tau \|\tilde{V}(\tau)\|$$

Hence the function $\tilde{Q}(\tau)$ is Lipschitz continuous with Lipschitz constant $\max_{[0,T]} \|\tilde{V}(\tau)\|^2$ that can be bounded by using the tightness of $\tilde{V}(\tau)$. Hence the tightness of $\tilde{Q}(\tau)$.

To prove the tightness of $\tilde{t}(\tau)$ we consider intercollision times

$$s_j = \hat{t}_{j+1} - \hat{t}_j = \hat{d}_j/\|v_j\|,$$

where d_j is the distance between the points of the jth and $(j+1)$st collisions and $\hat{d}_j = d_j 1_{j \leq \mathbf{kn}}$, for all $0 \leq j \leq M^{2/3}\bar{c}$. Note that $\|v_j\| \geq$ Const > 0 for all $j < \mathbf{kn}$, hence $s_j \leq$ Const. Let $\hat{L}_j$ equal $\bar{L}$ if $j < \mathbf{kn}$ and 0 otherwise.

Consider the function $d(x)$, $x \in \Omega$, equal to the distance between the positions of the light particle at the points x and $\mathcal{F}(x)$ (the distance between its successive collisions). In Section B.3 we prove the following:

PROPOSITION 6.14. *The function d belongs in our space $\mathfrak{R}$. The average $\mu_{Q,V}(d^k)$ is a Lipschitz continuous function of Q, V for $k \in \mathbb{N}$. In particular, we have*

$$\mu_{Q,V}(d) = \bar{L} + \mathcal{O}(\|V\|)$$

where $\bar{L}$ is the mean free path defined by (1.17).

Let $A(x) = d(x) - \mu_{Q,V}(d)$ and $B(x) = \mu_{Q,V}(d) - \bar{L}$. Then $d(x) = A(x) + \bar{L} + B(x)$ and, accordingly, $\hat{d}(x) = \hat{A}(x) + \hat{L}(x) + \hat{B}(x)$. Therefore

$$\tilde{t}(\tau) = \frac{1}{M^{1/3}} \sum_{j=0}^n \frac{\hat{A}(x_j)}{\|v_j\|}$$

$$+ \frac{1}{M^{1/3}} \sum_{j=0}^n \hat{L}(x_j)\left(\frac{1}{\|v_j\|} - 1\right) + \frac{1}{M^{1/3}} \sum_{j=0}^n \frac{\hat{B}(x_j)}{\|v_j\|}$$

$$= \tilde{t}_1(\tau) + \tilde{t}_2(\tau) + \tilde{t}_3(\tau)$$

where $n = M^{2/3}\tau$. The function $A(x)/\|v(x)\|$ satisfies the conditions of Proposition 6.1, in particular $\mu_{Q,V}(A/\|v\|) = 0$, hence $\tilde{t}_1(\tau)$ is tight due to Proposition 6.2 (a). Next

$$\frac{1}{\|v_j\|} - 1 = \frac{1 - \sqrt{1 - M\|V_j\|^2}}{\|v_j\|} = \mathcal{O}\left(M\|V_j\|^2\right) \tag{6.34}$$

To prove the equicontinuity of $\tilde{t}_2(\tau)$ we observe that

$$|\tilde{t}_2(\tau_2) - \tilde{t}_2(\tau_1)| \leq \text{Const} \frac{M^{2/3}|\tau_2 - \tau_1|}{M^{1/3}} \max_{n \leq \bar{c}M^{2/3}} \left(M\|V_n\|^2\right)$$
$$= |\tau_2 - \tau_1| \max_{\tau \leq \bar{c}} \|\tilde{V}(\tau)\|^2.$$

Hence, as before, the function $\tilde{t}_2(\tau)$ is Lipschitz continuous with Lipschitz constant $\max_{[0,T]} \|\tilde{V}(\tau)\|^2$ that can be bounded due to the tightness of $\tilde{V}(\tau)$. To prove the equicontinuity of $\tilde{t}_3(\tau)$ we use Proposition 6.14 and write

$$\left|\mu_{Q,V}(d) - \bar{L}\right| = \left|\mu_{Q,V}(d) - \mu_{Q,0}(d)\right| \leq \text{Const} \|V\|,$$

hence

$$|\tilde{t}_3(\tau_2) - \tilde{t}_3(\tau_1)| \leq \frac{\text{Const}}{M^{1/3}} |\tau_2 - \tau_1| \max_{\tau \leq \bar{c}} \|\tilde{V}(\tau)\|,$$

which is not only bounded due to the tightness of $\tilde{V}$, but can be made arbitrarily small.

Proposition 6.2 is proved. □

6.6. Second moment. Here we prove Proposition 6.4. We work in the context of Theorem 2, hence $\kappa_M = M^{-\delta}$ and $\mathbf{n} = M^{1/2-\delta}$. The context of Theorem 1 will be discussed in the next section.

Recall that $n = \varkappa M^{2/3}$. Our first step is to show that under the conditions of Proposition 6.1

$$\mathbb{E}_\ell(\hat{S}_n^2) = n\left[D_{\bar{Q},\bar{V}}(A) + g\right], \tag{6.35}$$

where $g \to 0$ as $M \to \infty$ and $\varkappa \to 0$ uniformly over all standard pairs with $\pi_1(\gamma) \subset \Upsilon^*_{\delta_2,a}$ and $\text{length}(\gamma) > M^{-100}$.

Indeed, by (6.31) and Lemma 6.12 (b) we have

$$\mathbb{E}_\ell(\hat{S}_n^2) = \mathbf{n} \sum_{k=0}^{n/\mathbf{n}} \mathbb{E}_\ell(\hat{D}_{Q_{k\mathbf{n}},V_{k\mathbf{n}}}) + \mathcal{O}\left(M^{1/3+2\delta}\right) + o(\mathbf{n}).$$

By Proposition 6.2, for most of the initial conditions the quantity

$$\max_{k < n/\mathbf{n}} \left\{\|Q_{k\mathbf{n}} - \bar{Q}\|, M^{2/3}\|V_{k\mathbf{n}} - \bar{V}\|\right\}$$

is small if $\varkappa$ is small, hence for most of the initial conditions $x \in \gamma$ we have $\mathbf{k}(x) \geq n/\mathbf{n}$ thus $\hat{D}_{Q_{k\mathbf{n}},V_{k\mathbf{n}}}(x) = D_{Q_{k\mathbf{n}},V_{k\mathbf{n}}}(A)$, and so we would only make small error if we replace $\hat{D}_{Q_{k\tau},V_{k\tau}}$ by $D_{\bar{Q},\bar{V}}(A)$ (note that

$D_{Q,V}(A)$ is a bounded and continuous function of Q, V on the domain $\text{dist}(Q, \partial \mathcal{D}) > \mathbf{r} + \delta_1$). Thus we obtain (6.35).

Now the parts (a)–(c) of Proposition 6.4 easily follow from Proposition 6.1 (a), (c), and (6.35). To prove (d), we write

$$\mathbb{E}_\ell(\hat{Q}_n - \bar{Q}) = \mathbb{E}_\ell \left(\sum_{j=0}^{n-1} s_j \hat{V}_j \right)$$

$$= \bar{V} \mathbb{E}_\ell \left(\sum_{j=0}^{n-1} s_j \right) + \mathbb{E}_\ell \left(\sum_{j=0}^{n-1} s_j (\hat{V}_j - \bar{V}) \right) = I + I\!I$$

where $s_j = \hat{t}_{j+1} - \hat{t}_j$ is the intercollision time. To estimate I we use the notation of the proof of Proposition 6.2 (b) and write:

$$\mathbb{E}_\ell \left(\sum s_j \right) = \mathbb{E}_\ell \left(s_j - \hat{L}_j / \|v_j\| \right) + \mathbb{E}_\ell \left(\hat{L}_j / \|v_j\| \right) = I_a + I_b$$

As we noted earlier, the function $s_j - \hat{L}_j / \|v_j\|$ satisfies the assumptions of Proposition 6.1, hence its part (a) implies $I_a = \mathcal{O}\left(M^{1/6+\delta}\right)$. By using (6.34) and Proposition 6.2 (b) we get

$$I_b = (1 + o_{\varkappa \to 0}(1)) \varkappa \bar{L} M^{2/3}.$$

Next, by the Cauchy-Schwartz inequality and Proposition 6.4 (b)

$$|I\!I| \leq \text{Const} \sum_j \sqrt{\mathbb{E}_\ell(\|\hat{V}_j - \bar{V}\|^2)}$$

$$\leq \text{Const} \sum_j \frac{j^{1/2}}{M}$$

$$\leq \text{Const} \frac{(\varkappa M^{2/3})^{3/2}}{M} \leq \text{Const}\, \varkappa^{3/2}.$$

This implies (d). To prove (e), we write, in a similar manner,

$$\mathbb{E}_\ell \left(\|\bar{Q}_n - \bar{Q}\|^2 \right) \leq 2 \|\bar{V}\|^2 \mathbb{E}_\ell \left(\left[\sum s_j \right]^2 \right) + 2 \mathbb{E}_\ell \left(\left\| \sum s_j (\hat{V}_j - \bar{V}) \right\|^2 \right)$$
$$= I + I\!I.$$

Then we have

$$|I| \leq \text{Const}\, \varkappa^2 \|\bar{V}\|^2 M^{4/3} = \mathcal{O}(\varkappa^2)$$

and
$$|I\!I| \leq \text{Const } \varkappa M^{2/3} \sum \mathbb{E}_\ell \left(s_j^2 \|\hat{V}_j - \bar{V}\|^2 \right)$$
$$\leq \text{Const } \varkappa M^{2/3} \sum \frac{j}{M^2}$$
$$\leq \text{Const } \varkappa M^{2/3} \frac{(\varkappa M^{2/3})^2}{M^2} = \mathcal{O}\left(\varkappa^3\right).$$

Proposition 6.4 is proven. □

6.7. Martingale property. To prove Proposition 6.5 we need to show that for every $m \geq 1$, all bounded and Lipschitz continuous functions $B_1, \ldots, B_m$ on the (Q, V) space, and all times $s_1 < s_2 < \cdots < s_m \leq \tau_1 < \tau_2$ we have

$$\mathbb{E}\left(\left[\prod_{i=1}^m B_i(\hat{\mathbf{Q}}(s_i), \hat{\mathbf{V}}(s_i)) \right] \left[\mathbf{M}(\tau_2) - \mathbf{M}(\tau_1) \right] \right) = 0.$$

where $\mathbb{E}$ denotes the expectation and
$$\mathbf{M}(\tau_2) - \mathbf{M}(\tau_1) = B(\hat{\mathbf{Q}}(\tau_2), \hat{\mathbf{V}}(\tau_2)) - B(\hat{\mathbf{Q}}(\tau_1), \hat{\mathbf{V}}(\tau_1))$$
$$- \int_{\tau_1}^{\tau_2} (\mathcal{L}B)(\hat{\mathbf{Q}}(s), \hat{\mathbf{V}}(s))\, ds$$

In other words, we have to show that
(6.36)
$$\mathbb{E}_{\max}\left(\left[\prod_{i=1}^m B_i\left(\tilde{Q}(s_i), \tilde{V}(s_i) \right) \right] \left[\beta_{J_2} - \beta_{J_1} - M^{-2/3} \sum_{j=J_1}^{J_2} \zeta_j \right] \right) \to 0$$

as $M \to \infty$, where
$$\beta_j = B\bigl(\hat{Q}_j, M^{2/3}\hat{V}_j\bigr), \qquad \zeta_j = \mathcal{L}B\bigl(\hat{Q}_j, M^{2/3}\hat{V}_j\bigr).$$
and
$$J_1 = M^{2/3}\tau_1, \qquad J_2 = M^{2/3}\tau_2$$

(see (6.25) for the definition of $\mathbb{E}_{\max}(\cdot)$ and note that B, $\mathcal{L}B$, and B_i are bounded and continuous functions). Lemma 6.11 allows us to eliminate the first factor in (6.36) and reduce it to

(6.37) $$\mathbb{E}_{\max}\left(\beta_J - \beta_0 - M^{-2/3} \sum_{j=0}^J \zeta_j \right) \to 0 \quad \text{as } M \to \infty$$

where $J = M^{2/3}(\tau_2 - \tau_1)$ (note that even if $s_m = \tau_1$, we can approximate
$$B_m((\tilde{Q}(s_m), \tilde{V}(s_m)) \approx B_m(\hat{Q}_{s_m M^{2/3} - M^\delta}, \hat{V}_{s_m M^{2/3} - M^\delta}),$$
$$B((\tilde{Q}(\tau_1), \tilde{V}(\tau_1)) \approx B(\hat{Q}_{\tau_1 M^{2/3}}, \hat{V}_{\tau_1 M^{2/3}}),$$

so Lemma 6.11 applies).

We will denote $\tau_2 - \tau_1$ by τ.

Next we prove (6.37). Given a small constant $\varkappa > 0$ and a large constant $R > 0$, we define $\hat{Q}', \hat{V}'$ similarly to $\hat{Q}, \hat{V}$ but with an additional stopping rule, defined in in the notation of Section 6.1: at any time moment k that is a multiple of $[\varkappa M^{2/3}/\mathbf{n}]$, we "remove from the circulation" all the standard pairs $\ell_{\alpha,k} = (\gamma_{\alpha,k}, \rho_{\alpha,k})$ where $\|V\| > M^{-2/3}R$ for some point $(Q, V) \in \pi_1(\gamma_{\alpha,k})$ (technically, we add the corresponding curve $\mathcal{F}^{-k\mathbf{n}}(\gamma_{\alpha,k})$ to the set I_k, see 6.1), and we do not change the construction of Section 6.1 for any time k that is not a multiple of $[\varkappa M^{2/3}/\mathbf{n}]$. Thus, the set I_k may get larger and $\mathbf{k}(x)$ may decrease, respectively. However, by Proposition 6.2 (b) we have, uniformly in $\varkappa$,

$$\sup_\ell \mathrm{mes}_\ell \left\{ (\hat{Q}', \hat{V}') \neq (\hat{Q}, \hat{V}) \right\} \to 0$$

as $R \to \infty, M \to \infty$, where the supremum is taken over all standard pairs $\ell = (\gamma, \rho)$ with $\mathrm{length}(\gamma) > M^{-100}$. Hence it is enough to show that for all large enough R

$$(6.38) \qquad \lim_{\varkappa \to 0} \lim_{M \to \infty} \mathbb{E}_{\max} \left(\beta'_J - \beta'_0 - M^{-2/3} \sum_{j=0}^{J} \zeta'_j \right) \to 0,$$

where

$$\beta'_j = B(\hat{Q}'_j, M^{2/3}\hat{V}'_j), \quad \beta'_0 = B(\bar{Q}, M^{2/3}\bar{V}),$$

$$\zeta'_j = \begin{cases} \mathcal{L}B(\hat{Q}'_j, M^{2/3}\hat{V}'_j), & \text{if } j \leq \mathbf{kn} \\ 0 & \text{otherwise} \end{cases}$$

$$J = M^{2/3}\tau.$$

(note that both expressions in parentheses in Eqs. (6.37) and (6.38) are uniformly bounded by a constant independent of R, because B has a compact support). To establish (6.38) it is enough to check that for all large R and uniformly in $k \leq \tau/\varkappa$

$$(6.39) \qquad \lim_{M \to \infty} \mathbb{E}_{\max} \left(\beta'_{(k+1)\mathbf{L}} - \beta'_{k\mathbf{L}} - M^{-2/3} \sum_{j=k\mathbf{L}}^{(k+1)\mathbf{L}} \zeta'_j \right) = o(\varkappa),$$

where $\mathbf{L} = \varkappa M^{2/3}$. To verify (6.39), we can assume, without loss of generality, that $k = 0$. Next we expand the function B into Taylor series about the point $(\bar{Q}, M^{2/3}\bar{V})$:

$$(6.40) \qquad \beta'_\mathbf{L} - \beta'_0 = \langle \nabla_Q B, dQ \rangle + \langle \nabla_V B, dV \rangle + \tfrac{1}{2}(dV)^T B_{VV} \, dV$$

$$(6.41) \qquad \qquad + \mathcal{O}\left(\|dQ\|^2 + \|dV\|^3 + \|dQ\|\|dV\|\right)$$

where $dQ = \hat{Q}'_\mathbf{L} - \bar{Q}$ and $dV = M^{2/3}(\hat{V}'_\mathbf{L} - \bar{V})$, and B_{VV} is a 2×2 matrix with components $\partial^2_{V_i,V_j} B$, $1 \leq i,j \leq 2$. We claim that

$$\mathbb{E}_\ell(\beta'_\mathbf{L} - \beta'_0) = M^{2/3}\bar{L}\langle \bar{V}, \nabla_Q B\rangle + \tfrac{1}{2}\sum_{i,j=1}^{2}\left(\bar{\sigma}^2_Q(\mathcal{A})\right)_{ij}\partial^2_{V_i,V_j}B + o(\varkappa)$$

(6.42) $$= (\mathcal{L}B)(\bar{Q}, M^{2/3}\bar{V})\varkappa + o(\varkappa)$$

for each standard pair $\ell = (\gamma, \rho)$ with length$(\gamma) > M^{-100}$ Indeed the terms in (6.40) are handled by Proposition 6.4(a), (b) and (d) whereas the terms in (6.41) are bounded as follows

$$\mathbb{E}(\|dQ\|^2) = \mathcal{O}(\varkappa^2)$$

by Proposition 6.4(e),

$$\mathbb{E}(\|dV\|^3) = \mathcal{O}(\varkappa^{3/2})$$

by Proposition 6.4(c) and Hölder inequality,

$$\mathbb{E}(\|dQ\|\|dV\|) = \mathcal{O}(\sqrt{\varkappa^2\varkappa}) = \mathcal{O}(\varkappa^{3/2})$$

by Proposition 6.4 and Cauchy-Schwartz (note that $\|\bar{V}\| < M^{-2/3}R$ due to our modified construction of $\hat{Q}'$ and $\hat{V}'$, hence Proposition 6.4 applies). On the other hand, by Proposition 6.2 (b)

$$\max_{j \leq \mathbf{L}}\left|\mathbb{E}_\ell\left[\mathcal{L}B(\hat{Q}'_j, M^{2/3}\hat{V}'_j) - \mathcal{L}B(\bar{Q}, M^{2/3}\bar{V})\right]\right| = o_{M\to\infty, \varkappa\to 0}(1),$$

hence

(6.43) $$\mathbb{E}_\ell\left(M^{-2/3}\sum_{j=0}^{\mathbf{L}}\zeta'_j\right) = \mathcal{L}B(\bar{Q}, \bar{V})\varkappa\left(1 + o_{M\to\infty, \varkappa\to 0}(1)\right).$$

Now (6.42) and (6.43) imply (6.39). Proposition 6.5 is proved. $\square$

6.8. Transition to continuous time. Here we prove Corollary 6.6. Pick a $\tau \in (0, \bar{c}\bar{L})$ and denote $t = M^{2/3}\tau$. For every $x \in \Omega$ choose n so that $t_n \leq t < t_{n+1}$. Then

$$\tilde{Q}_*(\tau) = \hat{Q}_n + \mathcal{O}(1/\sqrt{M})$$
$$= \hat{Q}_{[t/\bar{L}]} + (\hat{Q}_n - \hat{Q}_{[t/\bar{L}]}) + \mathcal{O}(1/\sqrt{M})$$

By Proposition 6.2 (b)

$$\text{mes}_\ell\left(\|\hat{Q}_n - \hat{Q}_{[t_n/\bar{L}]}\| > \max_{|n_1-n_2|<M^{1/3+\delta}}\|\hat{Q}_{n_1} - \hat{Q}_{n_2}\|\right) \to 0$$

as $M \to \infty$. By the tightness of $\tilde{V}(\tau)$

$$\mathrm{mes}_\ell \left(\max_{|n_1-n_2|<M^{1/3+\delta}} \|\hat{Q}_{n_1} - \hat{Q}_{n_2}\| > M^{-1/3+2\delta} \right) \to 0.$$

Combining these estimates gives

$$\Delta_Q := \mathrm{mes}_\ell \left(\sup_\tau \|\tilde{Q}_*(\tau) - \tilde{Q}(\tau/\bar{L})\| > \varepsilon \right) \to 0$$

as $M \to \infty$. We also claim that

(6.44) $$\Delta_V := \mathrm{mes}_\ell \left(\sup_\tau \|\tilde{V}_*(\tau) - \tilde{V}(\tau/\bar{L})\| > \varepsilon \right) \to 0$$

but this requires a slightly different argument. The tightness of $\tilde{V}(\tau)$ means that for any $\varepsilon > 0$ and $\varepsilon' > 0$ there is $\varepsilon'' > 0$ such that

$$\mathrm{mes}_\ell \left(\sup_{|n_1-n_2|<M^{2/3}\varepsilon''} \|\hat{V}_{n_1} - \hat{V}_{n_2}\| > M^{-2/3}\varepsilon \right) < \varepsilon'.$$

Hence, as before,

$$\Delta_V < \mathrm{mes}_\ell \left(\sup_{|n_1-n_2|<M^{1/3+\delta}} \|\hat{V}_{n_1} - \hat{V}_{n_2}\| > M^{-2/3}\varepsilon \right) + o(1)$$
$$< \varepsilon' + o(1)$$

as $M \to \infty$. The arbitrariness of ε' implies (6.44).

Thus each $(\mathbf{Q}_*, \mathbf{V}_*)$ can be obtained form the corresponding $(\hat{\mathbf{Q}}, \hat{\mathbf{V}})$ by the time change $\tau \to \tau/\bar{L}$. □

Remark. In the proof of Corollary 6.6 we used the tightness of $\tilde{t}(\tau)$, but it would be enough if the following function

(6.45) $$\tilde{t}_\Diamond(\tau) = M^{-1/3-\delta/2} \left[\hat{t}_{[\tau M^{2/3}]} - \bar{L} \min\{\tau M^{2/3}, kn\} \right]$$

was tight for some $\delta > 0$. We will refer to this observation in Chapter 8.

6.9. Uniqueness for stochastic differential equations.

Here we establish the uniqueness of solutions of (2.18) under the assumption that $\sigma_Q(\mathcal{A})$ satisfies (2.16).

There are two types of uniqueness for stochastic differential equations. *Pathwise uniqueness* means, in our terms, that given a Brownian motion $\mathbf{w}(\tau)$, any two solutions $(\mathbf{Q}_1(\tau), \mathbf{V}_1(\tau))$ and $(\mathbf{Q}_2(\tau), \mathbf{V}_2(\tau))$ such that $(\mathbf{Q}_1, \mathbf{V}_1)(0) = (\mathbf{Q}_2, \mathbf{V}_2)(0)$ coincide almost surely. *Uniqueness in distribution* means that any two solutions of the SDE have equal distributions provided their initial distributions coincide. We need the uniqueness in distribution, but according to [**76**, Section IX.1] it follows from the pathwise uniqueness, so we shall establish the later.

6. MOMENT ESTIMATES

Our argument follows [49, Section III]. Let $(\mathbf{Q}_1(\tau), \mathbf{V}_1(\tau))$ and $(\mathbf{Q}_2(\tau), \mathbf{V}_2(\tau))$ be two solutions with the same initial conditions. Denote
$$\Delta \mathbf{Q}(\tau) = \mathbf{Q}_1(\tau) - \mathbf{Q}_2(\tau), \quad \Delta \mathbf{V}(\tau) = \mathbf{V}_1(\tau) - \mathbf{V}_2(\tau)$$
We need to show that $(\Delta \mathbf{Q}, \Delta \mathbf{V})(\tau) \equiv 0$ with probability one. Given $k > 0$ let
$$\bar{\tau}_k = \sup\{\tau : \|\Delta \mathbf{Q}(\tau)\| < 0.1, \ \|\mathbf{V}_j(\tau)\| < k, \ j = 1, 2\}$$
and for every $\tau \geq 0$ we set $\tau_k = \min\{\tau, \bar{\tau}_k\}$. Let
$$a(\tau) = \mathbb{E}\left(\max_{s \leq \tau_k} \|\Delta \mathbf{V}(s)\|^2\right), \quad b(\tau) = \mathbb{E}\left(\max_{s \leq \tau_k} \|\Delta \mathbf{Q}(s)\|^2\right)$$
where $\mathbb{E}$ denotes the mean value. Since the coefficients of (2.18) are bounded due to our cutoffs, the functions $a(\tau)$ and $b(\tau)$ are continuous. Our goal is to establish that $a(\tau) = b(\tau) \equiv 0$ for each $k > 0$.

Observe that
$$\Delta \mathbf{V}(\tau) = \int_0^\tau \left[\sigma_{\mathbf{Q}_1(s)}(\mathcal{A}) - \sigma_{\mathbf{Q}_2(s)}(\mathcal{A})\right] d\mathbf{w}(s)$$
due to (2.18), hence $\Delta \mathbf{V}(\tau)$ is a martingale. By Doob's maximal inequality
$$a(\tau) \leq C_1 \mathbb{E}[\Delta \mathbf{V}(\tau_k)]^2$$
(here and on $C_i > 0$ are independent of $(\mathbf{Q}_1, \mathbf{V}_1)$ and $(\mathbf{Q}_2, \mathbf{V}_2)$). By L^2-isomorphism property of stochastic integration
$$a(\tau) \leq C_2 \mathbb{E} \int_0^{\tau_k} \left\|\sigma_{\mathbf{Q}_1(s)}(\mathcal{A}) - \sigma_{\mathbf{Q}_2(s)}(\mathcal{A})\right\|^2 ds$$
Now (2.16) yields
$$a(\tau) \leq C_3 \int_0^{\tau_k} \mathbb{E}\left(\|\Delta \mathbf{Q}(s)\|^2 \ln^2 \|\Delta \mathbf{Q}(s)\|^2\right) ds$$
Observe that the function $G(s) = s \ln^2 s$ is convex on the interval $0 \leq s \leq 0.1$, and $\|\Delta \mathbf{Q}(s)\| \leq 0.1$ for all $s \leq \tau_k$. Thus, Jensen's inequality yields

(6.46) $$a(\tau) \leq C_3 \int_0^{\tau_k} b(s) \ln^2 b(s) \, ds$$

On the other hand,
$$\|\Delta \mathbf{Q}(\tau)\|^2 = \left\|\int_0^{\tau_k} \Delta \mathbf{V}(s) \, ds\right\|^2 \leq C_4 \tau \int_0^{\tau_k} \|\Delta \mathbf{V}(s)\|^2 \, ds$$

hence

$$b(\tau) \leq C_5 \int_0^{\tau_k} a(s)\, ds \tag{6.47}$$

Our next goal is to show that (6.46) and (6.47), along with initial conditions $a(0) = b(0) = 0$, imply $a(\tau) = b(\tau) \equiv 0$. We use the following form of Gronwell inequality (see e.g. [**49**, Chapter III] for the proof of such results):

LEMMA 6.15. *Let f and g be monotone functions on a rectangle $R = [a_1, a_2] \times [b_1, b_2]$ and continuous functions $a(t)$ and $b(t)$ satisfy*

$$a(t) \leq \int_0^t f(a(s), b(s))\, ds, \qquad b(t) \leq \int_0^t g(a(s), b(s))\, ds$$

Let A and B be solutions of the differential equations

$$A' = f(A, B), \qquad B' = g(A, B)$$

If $(a(s), b(s)) \in R$ and $(A(s), B(s)) \in R$ for $0 \leq s \leq t$ and

$$a(0) \leq A(0), \qquad b(0) \leq B(0)$$

then

$$a(s) \leq A(s), \qquad b(s) \leq B(s)$$

for all $0 \leq s \leq t$.

This lemma (and the fact that $b < 0.01$) allows us to compare the functions $a(\tau)$ and $b(\tau)$ with the solutions of the differential equations

$$A' = C_4 B \ln^2 B, \qquad B' = C_5 A \tag{6.48}$$

with initial conditions $A(0) = B(0) = 0$. Our goal is to show that $A(\tau) = B(\tau) \equiv 0$ is the only nonnegative solution of the above initial value problem, i.e. there is no branching at $\tau = 0$.

Observe that (6.48) is a Hamiltonian-type system whose Hamiltonian

$$H = \tfrac{1}{2} C_5 A^2 - C_4 \int_0^B u \ln^2 u\, du$$
$$= \tfrac{1}{2} C_5 A^2 - C_4 \left(\tfrac{1}{2} B^2 \ln^2 B - \tfrac{1}{2} B^2 \ln B + \tfrac{1}{4} B^2 \right)$$

remains constant on all solutions (i.e. $H' \equiv 0$). On every solution originating at $(0,0)$, we have $H(\tau) \equiv 0$. Therefore, for small A, B we have $A \sim B|\ln B|$, hence

$$|B'| \leq C_6 B |\ln B|$$

6. MOMENT ESTIMATES

It remains to show that any such function B must be identically zero. Indeed, if $B_0 = B(\tau_0) > 0$ for some $\tau_0 > 0$, then

$$\tau_0 \geq C_6^{-1} \int_0^{B_0} \frac{dB}{B|\ln B|}$$

which is impossible because this integral diverges.

CHAPTER 7

Fast slow particle

Here we prove Theorem 1, which allows the slow particle (the disk) to move faster than Theorem 2 does. Our arguments are similar to those presented in Chapter 6, in fact now they are easier, because we only need to control the dynamics during time intervals $\mathcal{O}\left(M^{1/2}\right)$, instead of $\mathcal{O}\left(M^{2/3}\right)$.

Recall that for the proof of Theorem 1 we set κ_M to a small constant independent of M. Observe that Propositions 6.7 and 6.8, as well as Lemma 6.12, are applicable in the context of Theorem 1, but in the rest of Chapter 6 we assumed $\|\bar{V}\| \leq a M^{-2/3}$, which is not the case anymore. Instead of that, we will now assume that $M\|\bar{V}\|^2 \leq 1 - \delta_2$ (and $1 - \delta_2 > \chi$). We consider the dynamics up to $n \leq \bar{c}\sqrt{M}$ collisions, where $\bar{c} = c\sqrt{1-\chi^2}/\bar{L}$ and c is defined in Theorem 1 (note that $\sqrt{1-\chi^2}$ is the initial speed of the light particle, hence $\bar{L}/\sqrt{1-\chi^2}$ will approximate the mean intercollision time).

The following statement is analogous to Proposition 6.1.

PROPOSITION 7.1. *Assume the conditions of Proposition 6.1 but with a modified bound on the initial velocity: $M\|\bar{V}\|^2 \leq 1 - \delta_2$. Then, uniformly for $n \leq \bar{c}\sqrt{M}$, we have*

(a) $\mathbb{E}_\ell(\hat{S}_n) = \mathcal{O}\left(M^\delta\right).$

(b) $\mathbb{E}_\ell(\hat{S}_n^2) = \mathcal{O}(n).$

(c) $\mathbb{E}_\ell(\hat{S}_n^4) = \mathcal{O}(n^2).$

Proof. It is enough to divide $[0, \bar{c}]$ into intervals of length $\varkappa$ and apply Lemma 6.12 to each of them. □

Next we define certain continuous functions on the interval $[0, \bar{c}]$, in a way similar to (6.5)–(6.6), but with scaling factors specific to

Theorem 1:

$$\tilde{Q}(\tau) = M^{1/4}\left[\hat{Q}_{\tau M^{1/2}} - Q_0 - \frac{\bar{L}\min\{\tau M^{1/2}, \mathbf{kn}\}}{\sqrt{1-\chi^2}}V_0\right],$$

$$\tilde{V}(\tau) = M^{3/4}\left[\hat{V}_{\tau M^{1/2}} - V_0\right],$$

$$\tilde{S}(\tau) = M^{-1/4}\hat{S}_{\tau M^{1/2}},$$

$$\tilde{t}(\tau) = M^{-1/4}\left[\hat{t}_{[\tau M^{1/2}]} - \frac{\bar{L}\min\{\tau M^{1/2}, \mathbf{kn}\}}{\sqrt{1-\chi^2}}\right].$$

The next result is analogous to Proposition 6.2, and the proof only requires obvious modifications:

PROPOSITION 7.2. (a) *For every function* $A \in \mathfrak{R}$ *satisfying the assumptions of Proposition 6.1, the family of functions* $\tilde{S}(\tau)$ *is tight;* (b) *the families* $\tilde{Q}(\tau)$, $\tilde{V}(\tau)$, *and* $\tilde{t}(\tau)$ *are tight.*

The following result is similar to Proposition 6.4:

PROPOSITION 7.3. *Let* $\varkappa$ *be a small positive constant and* $n = \varkappa\sqrt{M}$. *The following estimates hold uniformly for all standard pairs* $\ell = (\gamma, \rho)$ *with* $\mathrm{length}(\gamma) > M^{-100}$ *and* $\pi_1(\gamma) \subset \Upsilon_{\delta_2}$, *and all* $(\bar{Q}, \bar{V}) \in \pi_1(\gamma)$:

(a) $\mathbb{E}_\ell(\hat{V}_n - \bar{V}) = \mathcal{O}(M^{-1+\delta})$.

(b) $\mathbb{E}_\ell\big((\hat{V}_n - \bar{V})(\hat{V}_n - \bar{V})^T\big) = \big(\bar{\sigma}^2_{\bar{Q},\bar{V}}(\mathcal{A}) + o_{\varkappa \to 0}(1)\big)\varkappa M^{-3/2}$.

(c) $\mathbb{E}_\ell(\|\hat{V}_n - \bar{V}\|^4) = \mathcal{O}(\varkappa^2 M^{-3})$.

(d) $\mathbb{E}_\ell(\hat{Q}_n - \bar{Q} - \hat{t}_n\bar{V}) = \mathcal{O}(\varkappa^{3/2} M^{-1/4})$.

In particular, if $\bar{V} = V_0 + uM^{-3/4}$, *for a* $u \in \mathbb{R}^2$, *then*

$$\mathbb{E}_\ell\left(\hat{Q}_n - \bar{Q} - \hat{t}_n V_0\right) = \frac{(1 + o_{\varkappa \to 0}(1))\bar{L}nu}{\sqrt{1-\chi^2}\,M^{3/4}} + \mathcal{O}(\varkappa^{3/2} M^{-1/4}).$$

(e) $\mathbb{E}_\ell\left(\|\hat{Q}_n - \bar{Q} - \hat{t}_n\bar{V}\|^2\right) = \mathcal{O}(\varkappa^3 M^{-1/2})$.

The proof goes along the same lines as that of Proposition 6.4. We only note that the proofs of parts (d) and (e) do not have to deal with the term $\sum_j s_j \bar{V}$ since it is included in $\hat{t}_n \bar{V}$, whereas the bound on $\sum_j s_j(V_j - \bar{V})$ is obtained exactly as before. Also note that the second estimate of part (d) follows from the first one and the fact that, by Lemma 6.12, $\mathbb{E}_\ell(\hat{t}_n) \sim n\bar{L}/\sqrt{1-\chi^2}$. □

The next statement is an analogue of Proposition 6.5:

PROPOSITION 7.4. *The function $\tilde{V}(\tau)$ weakly converges, as $M \to \infty$, to a Gaussian stochastic process $\tilde{\mathbf{V}}(\tau)$ with independent increments, zero mean, and the covariance matrix*

$$\mathrm{Cov}\,\tilde{\mathbf{V}}(\tau) = (1-\chi^2)\int_0^\tau \bar{\sigma}^2_{Q^\dagger\left(s\bar{L}/\sqrt{1-\chi^2}\right)}(\mathcal{A})\,ds.$$

Proof. Since $\tilde{V}(\tau)$ is tight, we only need to prove the convergence of finite dimensional distributions. Fix a $\tau < \bar{c}$ and choose $\varkappa \ll \tau$ so that $\tau/\varkappa \in \mathbb{N}$. Denote

$$R'_k = \hat{V}_{(k+1)\varkappa\sqrt{M}-M^\delta} - \hat{V}_{k\varkappa\sqrt{M}},$$

and

$$\tilde{V}'(\tau) = \sum_{k=0}^{\tau/\varkappa} R'_k.$$

Note that $\tilde{V}(\tau) - \tilde{V}'(\tau) = \mathcal{O}(M^{\delta-1/4}) \to 0$ as $M \to \infty$, hence the random processes $\tilde{V}'(\tau)$ and $\tilde{V}(\tau)$ must have the same finite dimensional limit distributions. By the continuity theorem, it is enough to prove the pointwise convergence of the corresponding characteristic functions, which we do next.

For every vector $\mathbf{z} \in \mathbb{R}^2$ we write Taylor expansion

$$\Phi_k(\mathbf{z}) := \exp\left(iM^{3/4}\langle \mathbf{z}, R'_k\rangle\right)$$
(7.1)
$$= 1 + iM^{3/4}\langle \mathbf{z}, R'_k\rangle - \tfrac{1}{2}M^{3/2}\langle \mathbf{z}, R'_k\rangle^2 + \mathcal{O}\left(M^{9/4}\langle \mathbf{z}, R'_k\rangle^3\right).$$

LEMMA 7.5. *For any standard pair $\ell = (\gamma, \rho)$ satisfying the conditions of Proposition 7.3 we have*

$$\mathbb{E}_\ell\big(\Phi_k(\mathbf{z})\big) = 1 - \tfrac{1}{2}(1-\chi^2)\,\varkappa\, \mathbf{z}^T D_k\,\mathbf{z} + o(\varkappa).$$

where

$$D_k = \bar{\sigma}^2_{Q^\dagger\left(k\varkappa\bar{L}/\sqrt{1-\chi^2}\right)}(\mathcal{A})$$

Proof. We apply Proposition 7.3 (a) and (b) to the linear and quadratic terms of (7.1), respectively, and bound the remainder term by the Hölder inequality:

$$\mathbb{E}_\ell\left(M^{9/4}|\langle \mathbf{z}, R'_k\rangle|^3\right) \leq M^{9/4}\left[\mathbb{E}_\ell\left(\langle \mathbf{z}, R'_k\rangle^4\right)\right]^{3/4}$$

and then use Proposition 7.3 (c). A delicate point here is to deal with the matrix $\bar{\sigma}^2_{Q,\bar{V}}(\mathcal{A})$ that comes from Proposition 7.3 (b). According to

Proposition 7.2, for most of the standard pairs $\ell = (\gamma, \rho)$

$$\bar{Q} = Q^\dagger\big(k\varkappa \bar{L}/\sqrt{1-\chi^2}\big) + \mathcal{O}\left(M^{-1/4+\delta}\right),$$
$$\bar{V} = V_0 + \mathcal{O}\left(M^{-3/4+\delta}\right),$$

where $k\varkappa$ is the time moment at which Proposition 7.3 (b) was applied. Since $\bar{\sigma}^2_{\bar{Q},\bar{V}}(\mathcal{A})$ is a bounded continuous function on the domain $\{(Q,V)\colon \mathrm{dist}(\bar{Q},\partial\mathcal{D}) > \mathbf{r}+\delta_2\}$, see Lemma A.10, we can replace $\bar{\sigma}^2_{\bar{Q},\bar{V}}(\mathcal{A})$ with

$$\bar{\sigma}^2_{Q^\dagger(k\varkappa\bar{L}/\sqrt{1-\chi^2}),V_0}(\mathcal{A}) = (1-\chi^2)D_k$$

the last equation follows from (1.15). □

Now by (7.1)

(7.2)
$$\begin{aligned}E_k(\mathbf{z})&:= \ln \mathbb{E}_\ell\big(\Phi_k(\mathbf{z})\big)\\ &= -\tfrac{1}{2}(1-\chi^2)\,\varkappa\, \mathbf{z}^T D_k\, \mathbf{z} + o(\varkappa).\end{aligned}$$

Let $0 \leq \tau' < \tau'' \leq \bar{c}$ be two moments of time such that $k' = \tau'/\varkappa \in \mathbb{N}$ and $k'' = \tau''/\varkappa \in \mathbb{N}$. Then

$$\begin{aligned}E_{\tau',\tau''} &:= \ln \mathbb{E}_\ell\left(\exp\left(iM^{3/4}\langle \mathbf{z}, \tilde{V}'(\tau'') - \tilde{V}'(\tau')\rangle\right)\right)\\ &= \ln \mathbb{E}_\ell\left(\prod_{k=k'}^{k''}\Phi_k(\mathbf{z})\right)\\ &= \sum_{k=k'}^{k''} E_k(\mathbf{z}) + o_{\varkappa\to 0}(1),\end{aligned}$$

where we used the same trick as in the proof of Lemma 6.11. By using (7.2) and letting $\varkappa \to 0$ we prove that for any $0 \leq \tau' < \tau'' \leq \bar{c}$

$$\lim_{M\to\infty} E_{\tau',\tau''} = -\frac{1-\chi^2}{2}\int_{\tau'}^{\tau''} \mathbf{z}^T\, \bar{\sigma}^2_{Q^\dagger(s\bar{L}/\sqrt{1-\chi^2})}(\mathcal{A})\,\mathbf{z}\,ds.$$

This shows that the increments of the limit process are Gaussian.

Next, let $0 \leq \tau_1 < \cdots < \tau_{m+1} \leq \bar{c}$ be arbitrary time moments and $\mathbf{z}_1,\ldots,\mathbf{z}_m \in \mathbb{R}^2$ arbitrary vectors. A similar computation as in Lemma 7.5 shows that the joint characteristic function of several increments

$$\mathbb{E}_\ell\left(\exp\left(iM^{3/4}\sum_{j=1}^m \langle \mathbf{z}_j, \tilde{V}'(\tau_{j+1}) - \tilde{V}'(\tau_j)\rangle\right)\right)$$

converges to

$$\exp\left(-\frac{1-\chi^2}{2}\sum_{j=1}^{m}\int_{\tau_j}^{\tau_{j+1}} \mathbf{z}_j^T \bar{\sigma}^2_{Q^\dagger\left(s\bar{L}/\sqrt{1-\chi^2}\right)}(\mathcal{A})\,\mathbf{z}_j\,ds\right).$$

as $M \to \infty$, hence the increments of the limiting process are independent. This completes the proof of Proposition 7.4. □

Lastly, the same argument as in the proof of Corollary 6.6 shows that the velocity function $\mathcal{V}(\tau)$ defined in Section 2.1 converges to the stochastic process $\mathbf{V}(\tau) = \tilde{\mathbf{V}}(\tau\sqrt{1-\chi^2}/\bar{L})$. The properties of $\mathbf{V}$ listed in Theorem 1 immediately follow from those of $\tilde{\mathbf{V}}$, which we proved above. The convergence of $\mathcal{Q}(\tau)$ to $\mathbf{Q}(\tau) = \int_0^\tau \mathbf{V}(s)\,ds$ follows from the fact that the integration is a continuous map on $C[0, \bar{c}\bar{L}]$. Theorem 1 is proved. □

CHAPTER 8

Small large particle

Here we prove Theorem 3, which requires the larger particle (the disk) shrink as $M \to \infty$.

First of all, the results of Chapter 6 apply to every $\mathbf{r} \in (0, \mathbf{r}_0)$, where $\mathbf{r}_0$ is a sufficiently small constant, and every time interval $(0, \bar{c})$. We now fix $\bar{c}_0 > 0$ and for each $\mathbf{r} \in (0, \mathbf{r}_0)$ apply those results to the time interval $(0, \bar{c})$ with

(8.1) $$\bar{c} = \bar{c}_\mathbf{r} = \mathbf{r}^{-1/3} \bar{c}_0$$

In other words, we consider a family of systems $\mathcal{F}_\mathbf{r} \colon \Omega_\mathbf{r} \to \Omega_\mathbf{r}$ (parameterized by $\mathbf{r}$), and for each of them obtain the results of Chapter 6 on the corresponding interval $(0, \bar{c}_\mathbf{r})$ with $\bar{c}_\mathbf{r}$ given by (8.1). Of course, all the $\mathcal{O}(\cdot)$ estimates in Chapter 6 will now implicitly depend on $\mathbf{r}$.

Next, for each $\mathbf{r}$ we define continuous functions on the interval $[0, \bar{c}_0]$, by the following rules that modify (6.5) and (6.45):

(8.2) $$\tilde{Q}(\tau) = \hat{Q}_{\tau \mathbf{r}^{-1/3} M^{2/3}}, \qquad \tilde{V}(\tau) = \mathbf{r}^{-1/3} M^{2/3} \hat{V}_{\tau \mathbf{r}^{-1/3} M^{2/3}},$$

and

(8.3) $$\tilde{t}_\Diamond(\tau) = \mathbf{r}^{1/6} M^{-1/3-\delta/2} \left[\hat{t}_{[\tau \mathbf{r}^{-1/3} M^{2/3}]} - \bar{L} \min\{\tau \mathbf{r}^{-1/3} M^{2/3}, \mathbf{k n}\} \right]$$

Now for each $\mathbf{r} \in (0, \mathbf{r}_0)$ we pick a function $A_\mathbf{r} \in \mathfrak{R}_\mathbf{r}$ (where $\mathfrak{R}_\mathbf{r}$ denotes the space $\mathfrak{R}$ defined in Section 3.3 corresponding to $\mathbf{r} > 0$), satisfying the assumptions of Proposition 6.1 with $\bar{c} = \mathbf{r}^{-1/3} \bar{c}_0$. Denote by $A = \{A_\mathbf{r}\}$ the family of just selected functions. Assume, additionally, that

(8.4) $$\mathfrak{S}_A := \sup_{0 < \mathbf{r} < \mathbf{r}_0} \max\{\|A_\mathbf{r}\|_\infty, \mathfrak{S}_{A_\mathbf{r}}\} < \infty$$

where $\mathfrak{S}_{A_\mathbf{r}}$ is computed according to (6.3). Now we define

(8.5) $$\tilde{S}(\tau) = \mathbf{r}^{1/6} M^{-1/3} \hat{S}_{\tau \mathbf{r}^{-1/3} M^{2/3}}.$$

PROPOSITION 8.1.
(a) *Given a family of functions $A = \{A_\mathbf{r}\}$ as above, there is a function $M_A(\mathbf{r})$ such that for $\mathbf{r} < \mathbf{r}_0$, $M > M_A(\mathbf{r})$, the family $\tilde{S}(\tau)$ is tight;*
(b) *There is a function $M_0(\mathbf{r})$ such that for $\mathbf{r} < \mathbf{r}_0$, $M > M_0(\mathbf{r})$, the families $\tilde{Q}(\tau)$, $\tilde{V}(\tau)$, and $\tilde{t}_\Diamond(\tau)$ are tight.*

The proof follows the same lines as that of Proposition 6.2, and we only discuss steps which require nontrivial modifications. The inequality (6.33) now implies (6.32) whenever $2^{-m} < \mathbf{r}^{4/21} M^{-8/21}$, cf. (8.4). For the case $2^{-m} < \mathbf{r}^{4/21} M^{-8/21}$ we need the following sharpened version of Lemma 6.13:

LEMMA 8.2. *Given a family $A = \{A_\mathbf{r}\}$ as above, there is a function $M_A(\mathbf{r})$ such that for all $\mathbf{r} < \mathbf{r}_0$, $M > M_A(\mathbf{r})$, all n_1, n_2 such that $|n_2 - n_1| > \mathbf{r}^{-1/7} M^{2/7}$ and all standard pairs $\ell = (\gamma, \rho)$ with $\mathrm{length}(\gamma) > M^{-100}$ and $\pi_1(\gamma) \subset \Upsilon^*_{\delta_2, a}$ we have*

$$\mathbb{E}_\ell\Big([\hat{S}_{n_2} - \hat{S}_{n_1}]^4\Big) \leq 3\mathfrak{S}_A^2 (n_2 - n_1)^2.$$

Proof. This bound follows from Lemma 6.12 (d) and the argument used in the proof of Proposition 6.1 (d). Note that the term $\mathcal{O}(\mathbf{n}^{1.9})$ in Lemma 6.12 (d) implicitly depends on $\mathbf{r}$, i.e. it is $< C(\mathbf{r})\,\mathbf{n}^{1.9}$, but we can always increase $M_0(\mathbf{r})$ so that $\mathbf{n}^{0.1} > M^{0.04} > C(\mathbf{r})/\mathfrak{S}_A^2$, hence $C(\mathbf{r})\,\mathbf{n}^{1.9} < \mathfrak{S}_A^2 \mathbf{n}^2$, as desired. □

Now the tightness of $\tilde{S}(\tau)$ follows due to (8.4).

To prove the tightness of $\tilde{V}(\tau)$, we need to modify the above argument slightly. Due to (1.9), (6.14) and Lemma 8.2

$$\mathbb{E}_\ell\Big([\hat{V}_{n_2} - \hat{V}_{n_1}]^4\Big) \leq \mathfrak{S}_A^2 M^{-4} (n_2 - n_1)^2 \leq \mathrm{Const}\, \mathbf{r}^2 M^{-4} (n_2 - n_1)^2,$$

where we used (2.20). Therefore,

$$\mathbb{E}_\ell\Big([\tilde{V}(\tau_2) - \tilde{V}(\tau_1)]^4\Big) \leq \mathrm{Const}\,(\tau_2 - \tau_1)^2,$$

which is sufficient to prove the equicontinuity of $\tilde{V}(\tau)$.

The tightness of $\tilde{Q}(\tau)$ and $\tilde{t}_\diamond(\tau)$ follows by the same argument as the one in the proof of Proposition 6.2. This involves the verification of (8.4) for the function $A(x) = d(x) - \mu_{Q,V}(d)$, which requires some effort. Fortunately, we can bypass this step by using the extra factor $M^{-\delta}$ included in the formula for $\tilde{t}_\diamond(\tau)$ and only verifying (8.4) for the function $A_\diamond(x) = M^{-\delta/2} A(x)$, which is much easier: it suffices to observe that $\mathfrak{S}_{A_\diamond} = M^{-\delta} \mathfrak{S}_A$ and choose $M_0(\mathbf{r})$ so that $M_0^\delta(\mathbf{r}) > \mathfrak{S}_A$ for every $\mathbf{r} < \mathbf{r}_0$. This gives $\mathfrak{S}_{A_\diamond} < 1$. □

Next, Lemma 6.12 (b) still holds, with some $g \to 0$, because we can use the same trick as above – increase $M_0(\mathbf{r})$, if necessary, to suppress the terms depending on $\mathbf{r}$. Having proved the tightness and Lemma 6.12 (b), we can derive estimates similar to those of Proposition 6.4.

The following statement is analogous to Proposition 7.4:

8. SMALL LARGE PARTICLE

PROPOSITION 8.3. *The function $\tilde{V}(\tau)$ weakly converges, as $\mathbf{r} \to 0$ and $M \to \infty$, $M > M_0(\mathbf{r})$, to the random process $\sigma_0 w_{\mathcal{D}}(\bar{L}\tau)$, in the notation of (2.25).*

The proof is similar to that of Proposition 7.4. A slight complication comes from the fact that, unlike Theorem 1, we have to stop the heavy particle when it comes too close to the border $\partial \mathcal{D}$. Thus we cannot argue the independence as before, since the increments depend on whether we have already stopped our particle or not. To overcome this complication, we let $\mathbf{w}(\tau)$ be the standard two dimensional Brownian motion (independent of our dynamical system) and define

$$V_n^{\diamond} = \begin{cases} V_n & \text{if } n \leq \mathbf{kn} \\ V_{\mathbf{kn}} + \mathbf{r}^{1/3} M^{-2/3} \sigma_0 [\mathbf{w}(\bar{L}n) - \mathbf{w}(\bar{L}\mathbf{kn})] & \text{otherwise} \end{cases}$$

(in other words, rather than terminating the velocity process once the particle comes too close to the border, we switch to an auxiliary Brownian motion). After this modification, the limiting process will have independent increments, and we can proceed as in the proof of Proposition 7.4. □

Lastly, the same argument as in the proof of Corollary 6.6 (see also the remark after it) shows that the limit of the functions $\mathcal{V}(\tau)$ defined in Section 2.3 and that of $\tilde{V}(\tau)$ above only differ by a time rescaling, $\tau \mapsto \tau/\bar{L}$, hence $\mathcal{V}(\tau)$ converges to the stochastic process $\sigma_0 w_{\mathcal{D}}(\tau)$, as claimed by Theorem 3. Finally, (2.26) follows by the fact that the integration is a continuous map on $C[0, \bar{c}\bar{L}]$. Theorem 3 is proved. □

CHAPTER 9

Open problems

Here we mention possible extensions of our results. More detailed discussion can be found in our survey [**22**].

9.1. Collisions of the massive disk with the wall. An important problem is to understand the behavior of $\bar{\sigma}_Q^2(\mathcal{A})$ as the disk $\mathcal{P}(Q)$ approaches the boundary of $\mathcal{D}$, since this would allow one to extend our results beyond the moment of the first collision of the heavy disk with the wall. It is natural to assume that this behavior should be controlled by the billiard dynamics in the domain where the heavy particle just touches $\partial \mathcal{D}$ at some point. This domain is still a dispersing billiard table, but two of its boundary components are tangent to each other (make two cusps). Therefore, one has to understand the mixing properties of dispersing billiards with cusps, which is a long standing open problem in billiard theory. There is a heuristic argument [**67**] that leads us to believe that discrete time correlations should decays as $\mathcal{O}(1/n)$, hence the diffusion matrix $\bar{\sigma}_Q^2(\mathcal{A})$ might be infinite or behave very irregularly. In any case, the dynamics in billiard tables with cusps appears to be quite delicate and requires further investigation. See [**27**] for recent results.

9.2. Longer time scales. In all the results of our paper, the velocity v of the light particle does not change significantly during the time intervals we consider, in fact its fluctuations converge to zero in probability as $M \to \infty$. On the basis of heuristic analysis of Section 1.3, we expect that v would experience changes of order one after $\mathcal{O}(M)$ collisions with the heavy disk. However, we are currently unable to treat such long intervals, since the error bounds we have in Proposition 3.3 and Corollary 3.4 would accumulate beyond $\mathcal{O}(1)$, so we need to improve upon this proposition in order to proceed further. We note that as the velocity of the light particle experiences changes of order one, the system starts approaching its natural equilibrium (its behavior is described by the invariant ergodic measure).

9.3. Stadia and the piston problem. The question of approaching a thermal equilibrium was recently considered by several authors for the piston model [**26**]. In that model a cubic container is divided into two compartments by a heavy insulating piston, and these compartments contain ideal gases at different temperatures. If the piston were infinitely heavy, it would not move and the temperature in each compartment would remain constant. However, if the mass of the piston is finite the temperatures would change slowly due to the energy and momenta exchanges between the particles and the piston. So far not much is known about the thermalization time needed for the temperatures to converge to a common limit value.

There is an obvious analogy between the motion of the piston and that of the heavy disk in our model. The dynamics of ideal gas particles in each compartment of in the piston model can be made hyperbolic by appropriate boundary conditions (say, let the container have a form of the Bunimovich stadium [**8**]). Then the methods of our paper could be used. Let us point out, however, that in our case the fluctuations about the averaged dynamics are diffusive, while in the piston case nondiffusive fluctuations may develop as follows. Some particles may move almost parallel to the piston bouncing back and forth between the flat walls of the container for a long time. If that happens on one side of the piston but not the other, the pressure balance will be broken, and the piston may be forced to move on a macroscopic scale.

9.4. Finitely many particles. The analysis of our paper extends without changes to systems with several heavy disks and one light particle. Of course, we need to prevent the disks from colliding with each other or the boundary of the table by restricting our analysis to a sufficiently short interval of time. Let us, for example, formulate an analogue of Theorem 2 in this situation (similar generalizations are possible for Theorem 1 and 3). Let k be the number of heavy disks which are initially at rest. Then after rescaling time by $M^{2/3}$, the velocity of the limiting process satisfy

$$d \begin{pmatrix} V_1 \\ \vdots \\ V_k \end{pmatrix} = \sigma_{Q_1 \ldots Q_k} \, d\mathbf{w}$$

where $\mathbf{w}$ is a standard $2k$-dimensional Brownian motion. Notice that even though the heavy disks are not allowed to approach each other, each one "feels" the presence of the others through the diffusion matrix $\sigma_{Q_1 \ldots Q_k}$ which depends on the positions of all the disks.

In order to extend our results to systems with several light particles, one needs to generalize Proposition 3.3. Here we have two possibilities. One is to work with a discrete time dynamics, then the multiparticle system is a semidispersing billiard in a higher dimensional space. Very little is known about mixing rates in such systems, see some results in [18]. Alternatively, we may work directly with a continuous time system, and in this case we get a direct product of 2D billiards. This would require obtaining the bounds on continuous time correlation functions, which should be possible in view of recent results [17, 65].

9.5. Growing number of particles. A more realistic model of Brownian motion consists of one heavy disk and many light particles, whose number grows with M. It is also quite reasonable to make the size of the heavy disk decrease as M grows. Let the diameter of the disk be $\mathbf{r} = M^{-\alpha}$ for a small $\alpha > 0$ and the number of light particles $N = M^{\beta}$ for a small $\beta > 0$. Since, in view of (1.2), the heavy disk "remembers" only the last $\mathcal{O}(M)$ collisions, it is natural to assume that its velocity will be of order $\sqrt{M}/M = 1/\sqrt{M}$. Hence it covers a distance of order one during a time interval of order $\sqrt{M}$. Let $\tau = t\sqrt{M}$. According to the calculations of Chapter 8, the expected number of collisions during this time interval is of order

$$N\sqrt{M}\mathbf{r} = M^{1/2+\alpha-\beta}.$$

On the other hand, (1.2) tells that $\mathcal{O}(M)$ is a critical number of collisions. Hence the following conjecture seems reasonable:

CONJECTURE 9.1. *Suppose that the initial state of each light particle is chosen independently, so that the position and velocity direction are uniformly distributed and the speed has an initial distribution with smooth density $\rho_0(v)$. Denote $a_j(\rho) = \int |v|^j \rho(v)\, dv$. Then the limiting process $\mathbf{Q}(\tau)$ is*
  (a) *straight motion if $\beta = \alpha + 1/2 - \epsilon$,*
  (b) *the integral of an Ornstein-Uhlenbeck process*

(9.1) $$d\mathbf{Q} = \mathbf{V}\, d\tau, \qquad d\mathbf{V} = -\nu \mathbf{V}\, d\tau + \sigma\, d\mathbf{w}$$

where

$$\nu = c_1 a_1(\rho_0), \qquad \sigma^2 = c_2 a_3(\rho_0)$$

if $\beta = \alpha + 1/2$
  (c) *a Brownian motion if $\beta = \alpha + 1/2 + \epsilon$.*

The justification of this conjecture is straightforward. In fact, part (a) is in direct analogy with Theorem 1. There are too few collisions to produce significant changes of the velocity of the massive disk. Part (b) is similar to Theorem 2, with one notable difference: in Theorem 2, there is no drift for the velocity of the disk since the number of collisions was too small for the factor $\frac{M-1}{M+1}$ in (1.2) to take effect. Under the setting of the above conjecture, it is this factor that determines the drift of the Ornstein-Uhlenbeck process. The factor a_1 in the drift term comes from the fact that the number of collisions of the massive disk with any given particle is proportional to the speed of that particle. The reason for the factor a_3 in the diffusion term is explained before Theorem 1 (see also [36]). Also, observe that in the two particle model treated in this paper, the velocity of the massive disk has a maximal value, $\frac{1}{\sqrt{M}}$, hence when it gets close to this value it is more likely to decrease than to increase. In this sense, we have a "superdrift" in the two particle model. Finally, in the case (c) the Ornstein-Uhlenbeck regime should take effect on time intervals which are much shorter than τ, hence (c) is quite natural in view of the fact that Ornstein-Uhlenbeck process satisfies the central limit theorem.

We believe that the cases of several light particles of the previous subsection and a growing number of particles discussed here are similar, on a technical level. Indeed, our arguments are based on the estimation of the first four moments. For arbitrary many particles, the computation of the fourth moment contains only the contribution of all 4-tuples of collisions, but each 4-tuple involves at most four different light particles, hence an extension of Proposition 3.3 to only four light particles should be enough for the study of systems with arbitrary many particles. In the case of a growing number of particles, there is also an additional complication because there are, inevitably, slow particles for which there is not enough time for mixing to take effect. However, we expect the contribution of those particles be small, since their collisions with the heavy disk will result in relatively small changes of the velocity of the latter (cf. also [36]).

9.6. Particles of positive size. The results of our paper obviously remain valid if the light particle has a positive diameter, $2r_0$, which is smaller than the shortest distance between scatterers $\mathbb{B}_i$. Indeed, that model can be reduced to ours by enlarging the massive disk and all the scatters by r_0. However, a model of several light particles of positive diameter becomes more interesting, since the particles can interact with each other. To fix our ideas, consider the situation of the previous subsection with $\beta = \alpha + 1/2$ but now let us assume that instead of $r_0 = 0$ we have $r_0 = M^{-\gamma}$. Then each light particle is expected to collide with $Nr_0\sqrt{M}$ other particles. Since momentum transferred during each collision is of order 1, now an interesting scaling regime is $Nr_0\sqrt{M} \sim 1$. In this case we can expect ρ to change according to the kinetic theory. Thus the following statement seems reasonable.

9. OPEN PROBLEMS

CONJECTURE 9.2. *The limiting process satisfies*

$$(9.2) \qquad d\mathbf{Q} = \mathbf{V}\,d\tau, \qquad d\mathbf{V} = -\nu(\tau)\,\mathbf{V}\,d\tau + \sigma(\tau)\,d\mathbf{w}$$

where

(a) $\nu = c_1 a_1(\rho_0)$, $\sigma^2 = c_2 a_3(\rho_0)$ *if* $\gamma > \alpha + 1$,

(b) $\nu = c_1 a_1(\rho_\tau)$, $\sigma^2 = c_2 a_3(\rho_\tau)$, *and ρ_t satisfies the homogeneous Boltzmann equation*

$$\frac{d\rho_\tau}{d\tau} = Q(\rho_\tau, \rho_\tau),$$

where Q is the Boltzmann collision kernel, if $\gamma = \alpha + 1$,

(c) $\nu = c_1 a_1(\rho_{\text{Max}}(a_2(\rho_0)))$ *and* $\sigma^2 = c_2 a_3(\rho_{\text{Max}}(a_2(\rho_0)))$, *where $\rho_{\text{Max}}(a_2(\rho_0))$ is the Maxwellian distribution with the same second moment as ρ_0, if $\gamma < \alpha + 1$.*

Acknowledgment. We are deeply indebted to Ya. G. Sinai who proposed to us many of the problems discussed in this paper. He outlined a general strategy of the proof of Theorem 1, suggested the main idea of the proof of Proposition 5.12, and we benefited from long conversations with him about the entire work. This work was done when the authors, in turn, stayed at the Institute for Advanced Study. We are grateful to the institute staff and especially to our host T. Spencer for the excellent working conditions. D. Dolgopyat would like to thank D. Ruelle for explaining to him the results of [**78**]. A local version of these results constitute an important tool in this paper. We thank Peter Balint for his comments on the preliminary version of this paper. N. Chernov was partially supported by NSF grants DMS-9729992, DMS-0098788. D. Dolgopyat was partially supported by NSF grant DMS-0245359, the Sloan Fellowship and IPST.

APPENDIX A

Statistical properties of dispersing billiards

Throughout the paper, we have made an extensive use of statistical properties of dispersing billiards obtained recently in [11, 18, 96]. On several occasions, though, those results were insufficient for our purposes, and we needed to extend or sharpen them. Here we adjust the arguments of [11, 18, 96] to obtain the results we need. The reader is advised to consult those papers and a recent book [28] for relevant details.

A.1. Decay of correlations: overview. To fix our notation, let $\mathcal{D} = \mathbb{T}^2 \setminus \cup_{i=0}^{r} \mathbb{B}_i$ be a dispersing billiard table, where $\mathbb{B}_0, \mathbb{B}_1, \ldots, \mathbb{B}_r$ are open convex scatterers with C^3 smooth boundaries and disjoint closures (the scatterer $\mathbb{B}_0$ will play a special role, it corresponds to the disk $\mathcal{P}(Q)$ in our main model). Denote by $\Omega_\mathcal{D} = \partial \mathcal{D} \times [-\pi/2, \pi/2]$ the collision space, $\mathcal{F}_\mathcal{D} \colon \Omega_\mathcal{D} \to \Omega_\mathcal{D}$ the collision map, and $\mu_\mathcal{D}$ the corresponding invariant measure.

Assume that the horizon is finite, i.e. the free path between collisions is bounded by $L_{\max} < \infty$. In this case for every $k \geq 1$ the map $\mathcal{F}_\mathcal{D}^k$ is discontinuous on a set $\mathcal{S}_k \subset \Omega_\mathcal{D}$, which is a finite union of smooth compact curves. The complement $\Omega_\mathcal{D} \setminus \mathcal{S}_k$ is a finite union of open domains which we denote by $\Omega_{\mathcal{D},k,j}$, $1 \leq j \leq J_k$.

Now let $\mathcal{H}_{k,\eta}$ denote the space of functions on $\Omega_\mathcal{D}$ which are Hölder continuous with exponent η on each domain $\Omega_{\mathcal{D},k,j}$, $1 \leq j \leq J_k$:

$$f \in \mathcal{H}_{k,\eta} \Leftrightarrow \exists K_f : \forall j \in [1, J_k] \ \forall x, y \in \Omega_{\mathcal{D},k,j}$$
$$|f(x) - f(y)| \leq K_f \left[\mathrm{dist}(x,y)\right]^\eta.$$

One of the central results in the theory of dispersing billiards is

PROPOSITION A.1 (Exponential decay of correlations [96]). *For every $\eta \in (0,1]$ and $k \geq 1$ there is a $\theta_{k,\eta} \in (0,1)$ such that for all $f, g \in \mathcal{H}_{k,\eta}$ and $n \in \mathbb{Z}$*

(A.1) $$\left| \mu_\mathcal{D}(f \cdot (g \circ \mathcal{F}_\mathcal{D}^n)) - \mu_\mathcal{D}(f) \mu_\mathcal{D}(g) \right| \leq C_{f,g} \theta_{k,\eta}^{|n|}$$

where

(A.2) $$C_{f,g} = C_0 (K_f + \|f\|_\infty)(K_g + \|g\|_\infty)$$

and $C_0 = C_0(\mathcal{D}) > 0$ is a constant.

The exponential bound (A.1) is stated and proved in [18, 96]. The formula (A.2), which we also need for our purposes, is not explicitly derived there, but it follows from the estimates on pages 608–609 of [96].

The arguments in [96] can be used to derive the following analogue of our Proposition 3.3:

PROPOSITION A.2 (Equidistribution for billiards). *For every $\eta \in (0,1]$ and $k \geq 1$ there is a $\theta_{k,\eta} \in (0,1)$ such that for any $f \in \mathcal{H}_{k,\eta}$ and any standard pair consisting of an H-curve $W \subset \Omega_\mathcal{D}$ and a smooth probability measure ν on it we have*

$$(\text{A.3}) \qquad \left| \int_W f \circ \mathcal{F}_\mathcal{D}^n \, d\nu - \mu_\mathcal{D}(f) \right| \leq C_f \theta_{k,\eta}^n \qquad \forall n \geq K |\ln|W||$$

where $C_f = C_0(K_f + \|f\|_\infty)$ and $C_0, K > 0$ are constants. In addition, by time reversibility, a similar property holds for stable curves and negative iterations of $\mathcal{F}_\mathcal{D}$.

Below we sketch alternative proofs of both Propositions A.1 and A.2 using a 'coupling method' [7, 97]. This is done in order to make our presentation self-contained, as well as to emphasize the central role played by shadowing-type arguments in the whole theory.

Since the rest of this subsection deals with a fixed domain, we drop $\mathcal{D}$ in $\mathcal{F}_\mathcal{D}$. First we derive Proposition A.1 from A.2. We may assume that $\mu(g) = 0$ (otherwise we replace g with $g - \mu(g)$).

Let $\mathcal{G} = \{\gamma_\alpha\}$ be a smooth foliation of $\Omega_\mathcal{D}$ by H-curves on which standard pairs can be defined. (Sufficiently smooth H-curves will do, alternatively such foliations are constructed in [20].) Denote by $\mathcal{G}' = \{\gamma'_\beta\}$ the foliation of $\Omega_\mathcal{D}$ into the H-components of the sets $\mathcal{F}^{n/2}(\gamma_\alpha)$, $\gamma_\alpha \in \mathcal{G}$. For every curve $\gamma'_\beta \in \mathcal{G}'$ its preimage $\mathcal{F}^{-n/2}(\gamma'_\beta)$ has length smaller than $C\vartheta^{n/2}$, where $\vartheta^{-1} > 1$ denotes the minimal expansion factor of unstable curve, cf. (4.8). Hence we can approximate the function f by a constant function on every curve $\mathcal{F}^{-n/2}(\gamma'_\beta)$, $\gamma'_\beta \in \mathcal{G}'$, and this approximation results in an error term $\mathcal{O}(\|g\|_\infty K_f \vartheta^{nn/2})$ (for all $n/2 > k$). Then we apply (A.3) to $n/2$ iterations of $\mathcal{F}$, the function g and every curve $\gamma'_\beta \in \mathcal{G}'$ whose length is at least $e^{-n/2K}$, and obtain a bound $\|f\|_\infty (K_g + \|g\|_\infty) \theta_{k,\eta}^{n/2}$. Lastly, the total measure of the curves $\gamma'_\beta \in \mathcal{G}'$ whose length is shorter than $e^{-n/2K}$ is $\mathcal{O}(e^{-n/2K})$ due to Lemma 4.10 (b), so their contribution will be bounded by $\mathcal{O}(\|f\|_\infty \|g\|_\infty e^{-n/2K})$. Thus Proposition A.1 follows. □

Next we prove Proposition A.2 in several steps.

Step 1. We may assume that W is long enough, i.e. $|W|$ is bounded away from zero, otherwise we apply Lemma 4.10 (c) to transform W into H-components of length $\geq \varepsilon_0$. Hence we assume that $|W| \geq \varepsilon_0$ (in this case (A.3) will hold for all $n \geq 1$). We will say that an H-curve W is *long* if $|W| \geq \varepsilon_0$.

Next, to establish (A.3) it is enough to show that the distribution of the image of $\mathcal{F}^n W$ is almost independent of W, that is

$$(A.4) \qquad \left| \int_{W_1} f \circ \mathcal{F}^n \, d\nu_1 - \int_{W_2} f \circ \mathcal{F}^n \, d\nu_2 \right| \leq C_f \theta_{k,\eta}^n$$

where (W_i, ν_i) satisfy the assumptions of Proposition A.2, and both W_1, W_2 are long. We will prove a (slightly) more general fact:

$$(A.5) \qquad \left| \int_M f \circ \mathcal{F}^n \, d\mu_1 - \int_M f \circ \mathcal{F}^n \, d\mu_2 \right| \leq C_f \theta_{k,\eta}^n$$

where μ_1, μ_2 are measures of the form

$$\mu_i = \int \mathrm{mes}_{\ell_\alpha} d\lambda_i(\alpha)$$

with $\mathcal{G} = \{\ell_\alpha\}$ being some family of standard pairs and λ_i factor measures on $\mathcal{G}$ satisfying

$$(A.6) \qquad \lambda_i\big(\mathrm{length}(\gamma_\alpha) \leq \varepsilon\big) \leq \mathrm{Const}\,\varepsilon.$$

We will say that a family of standard pairs with a factor measure λ_i is *proper* if it satisfies (A.6).

Note that (A.5) implies Proposition A.2 if we set λ_1 to an atomic measure (concentrated on a single long H-curve) and $\mu_2 = \mu_\mathcal{D}$, as $\mu_\mathcal{D}$ satisfies (A.6) by our discussion in Section 3.3 and [20].

Step 2. The proof of (A.4) will be accomplished by the so called coupling algorithm developed in [97]. Its main idea is to divide $\mathcal{F}^n W_1$ and $\mathcal{F}^n W_2$ into pieces, which can be paired so that the elements of each pair are close to each other (we used a similar idea to prove Proposition 3.2, but there we coupled the images of the same curve under different maps). However, since the expansion is not uniform in different regions of $\Omega_\mathcal{D}$, some pieces of $\mathcal{F}^n W_i$ may carry more weight than others, so we may have to couple a heavy piece with several light ones. This can be done by splitting a heavy piece into several 'thinner' curves, each coupled to a different partner. It is actually convenient to split each curve W_i into uncountable many 'fibers'. Namely, given a standard pair (W, ν), we consider $Y = W \times [0, 1]$ and equip Y with a probability measure

(A.7) $$dm(x, t) = d\nu(x)\, dt = \rho(x)\, dx\, dt$$

where $\rho(x)$ is the density of ν and $0 \leq t \leq 1$. We call Y a *rectangle* with *base* W. The map $\mathcal{F}^n$ can be naturally defined on Y by $\mathcal{F}^n(x, t) = (\mathcal{F}^n x, t)$ and the function f by $f(x, t) = f(x)$.

The coupling method developed in [97] will give us the following:

LEMMA A.3. *Let W_1 and W_2 be two long H-curves, and Y_1 and Y_2 the corresponding rectangles. Then there exist a measure preserving map (coupling map) $\xi \colon Y_1 \to Y_2$ and a function $R \colon Y_1 \to \mathbb{N}$ such that*

(A) *For all $(x, t) \in Y_1$ and $\xi(x, t) = (y, s) \in Y_2$ and all $n > R(x, t)$ the points $\mathcal{F}^n(x)$ and $\mathcal{F}^n(y)$ lie on the same stable manifold in the same connected component of $\Omega_\mathcal{D} \setminus \mathcal{S}_{-n+R(x,t)}$; in particular*

$$\mathrm{dist}(\mathcal{F}^n(x), \mathcal{F}^n(y)) \leq C\theta^{n-R(x,t)}$$

where $C > 0$ and $\theta < 1$ are constants.

(B) *For all n we have $m_1\big((x, t) \colon R(x, t) > n\big) \leq C\theta^n$.*

We postpone the proof untill step 3 and first derive (A.4) from Lemma A.3:

$$\Delta := \int_{W_1} f \circ \mathcal{F}^n \, d\nu_1 - \int_{W_2} f \circ \mathcal{F}^n \, d\nu_2$$

$$= \int_{Y_1} f(\mathcal{F}^n(x, t))\, dm_1 - \int_{Y_2} f(\mathcal{F}^n(y, s))\, dm_2$$

$$= \int_{Y_1} \big[f(\mathcal{F}^n(x, t)) - f(\mathcal{F}^n(\xi(x, t)))\big]\, dm_1.$$

The last integral can be decomposed as

$$\int_{Y_1} [\ldots] = \int_{R > n/2} [\ldots] + \int_{R \leq n/2} [\ldots] = I + I\!I,$$

and it is easy to see that $|I| \leq 2C\|f\|_\infty \theta^{n/2}$ and $|I\!I| \leq \mathrm{Const}\, K_f \theta^{n\eta/2}$. □

Step 3. Here we begin the proof of Lemma A.3. It is fairly long and technical; we describe all the major steps here, but a little more detailed presentation can be found in [20, Appendix].

First we construct a special family of stable manifolds that will be used to 'couple' points of Y_1 and Y_2. Let $\tilde{W} \subset \Omega_\mathcal{D}$ be an H-curve and $\kappa > 0$; define

$$\tilde{W}_\kappa = \tilde{W} \setminus \cup_{n \geq 0} \mathcal{F}^{-n} \mathcal{U}_{\kappa \vartheta^n}(\mathcal{S}_1)$$

where $\mathcal{U}_\varepsilon(\mathcal{S}_1)$ denotes the ε-neighborhood of $\mathcal{S}_1$. It is standard that through every point $x \in \tilde{W}_\kappa$ there is a stable manifold W_x^s extending at least the distance κ on both sides of $\tilde{W}$. We denote this family of stable manifolds by $\mathcal{G}_\kappa^s(\tilde{W})$.

Furthermore, $|\tilde{W} \setminus \cup_{\kappa>0}\tilde{W}_\kappa| = 0$. Hence by reducing $\tilde{W}$ we can ensure that, given any $D, \delta > 0$, we can find a curve $\tilde{W}$ and $\kappa > 0$ such that

(A.8) $\qquad \kappa > D|\tilde{W}| \quad$ and $\quad |\tilde{W}_\kappa|/|\tilde{W}| > 1 - \delta$.

Moreover, for every $x \in \tilde{W}_\kappa$ the set of points $y \in W_x^s$ such that the unstable manifold W_y^u intersects all the stable manifolds $W^s \in \mathcal{G}_\kappa^s(\tilde{W})$ has positive Lebesgue measure on W_x^s. For the rest of this section, we fix a small $\delta > 0$, such a curve $\tilde{W}$, the family $\mathcal{G}^s = \mathcal{G}_\kappa^s(\tilde{W})$, and denote their union by $\Lambda^s = \cup_{\mathcal{G}^s} W^s$. We will say that an H-curve W *fully crosses* Λ^s if it intersects *all* the stable manifolds $W^s \in \mathcal{G}^s$. We note that if D is large enough then the first inequality in (A.8) guarantees that any sufficiently long H-curve W that satisfies $\mathrm{dist}(W, \tilde{W}) < |\tilde{W}|$ will fully cross Λ^s (because the 'height' of Λ^s is much larger than its 'length'). Observe that $\tilde{W}_\kappa = \tilde{W} \cap \Lambda^s$. For any H-curve W fully crossing Λ^s we set $W_\kappa := W \cap \Lambda^s$.

Next, for any standard pair $\ell = (\gamma, \rho)$ and any $n \geq 0$ denote by $\gamma_{n,i}$ the H-components of $\mathcal{F}^n(\gamma)$ that fully cross Λ^s and put

(A.9) $\qquad \gamma_{n,*} = \cup_i \mathcal{F}^{-n}(\gamma_{n,i} \cap \Lambda^s)$.

We claim that there are constants $n_0 \geq 1$ and $d_0 > 0$ such that for any long standard pair (i.e. $|\gamma| \geq \varepsilon_0$) and any $n \geq n_0$ we have

(A.10) $\qquad \mathrm{mes}_\ell(\gamma_{n,*}) \geq d_0$.

This follows from the mixing property of $\mathcal{F}$ and the compactness of the set of long H-curves, the proof of (A.10) is essentially given in [**11**, Theorem 3.13].

Now let $\ell = (\gamma, \rho)$ be a standard pair such that γ fully crosses Λ^s, then $\gamma_\kappa = \gamma \cap \Lambda^s$ is a Cantor set on γ, and its complement $\gamma \setminus \gamma_\kappa$ consists of infinitely many intervals; we call them *gaps in* γ_κ. These gaps naturally correspond to the intervals of $\tilde{W} \setminus \tilde{W}_\kappa$ (gaps in $\tilde{W}_\kappa$), which are created by the removal of the $\mathcal{F}^{-n}$-images of the $c\vartheta^n$-neighborhoods of $\mathcal{S}_1$ from $\tilde{W}$. We call n the *rank* of the corresponding gap (if a gap is made by several overlapping intervals with different n's, then its rank is the smallest such n).

If a gap $\tilde{V} \subset \tilde{W} \setminus \tilde{W}_\kappa$ has rank n, then $\mathcal{F}^n(\tilde{V})$ will have length $\geq c\vartheta^n$. It corresponds to a gap $V \subset \gamma \setminus \gamma_\kappa$, to which we also assign rank n; observe that $\mathcal{F}^n(V)$ lies in the ε-vicinity of $\mathcal{F}^n(\tilde{V})$ with some $\varepsilon \ll \vartheta^n$, hence $\mathcal{F}^n(V)$ has length $\geq \frac{1}{2}c\vartheta^n$. Then the set $\mathcal{F}^{n(1+\beta_3|\ln \vartheta|)}(V)$, equipped with the image of the conditional measure $\mathrm{mes}_V = \mathrm{mes}_\ell(\cdot|_V)$ on V, will be a proper family of standard pairs, in the sense of (A.6), as it follows from Lemma 4.10 (b). Accordingly, we define a 'recovery time' function $r_\gamma(x)$ on $\gamma \setminus \gamma_\kappa$ by setting $r_\gamma(x) = n(1+\beta_3|\ln \vartheta|)$, where n is the rank of the gap containing the point x (note that the function $r_\gamma(x)$ is constant on every gap). Lemma 4.10 (b) implies that for some $\theta < 1$ and all $n > 0$

$$(\mathrm{A.11}) \qquad \mathrm{mes}_\ell(x \in \gamma \setminus \gamma_\kappa \colon r_\gamma(x) > n)/\mathrm{mes}_\ell(\gamma \setminus \gamma_\kappa) \leq \mathrm{Const}\, \theta^n.$$

Next, let $s_\ell(x)$ be another function on $\gamma \setminus \gamma_\kappa$ that is constant on every gap and such that $s_\ell(x) \geq r_\gamma(x) + n_0$. Then $\mathrm{mes}_V(V_{s_\ell(V),*}) \geq d_0$ for each gap $V \subset \gamma \setminus \gamma_\kappa$, in the notation of (A.9). We call s_ℓ a 'stopping time' function.

LEMMA A.4. *We can define the stopping time function $s_\ell(x)$ on $\gamma \setminus \gamma_\kappa$ so that for all $n \geq 1$*

$$(\mathrm{A.12}) \qquad \mathrm{mes}_\ell(x \in \gamma \setminus \gamma_\kappa \colon s_\ell(x) = n)/\mathrm{mes}_\ell(\gamma \setminus \gamma_\kappa) = q_n,$$

where $\{q_n\}$ is a sequence satisfying

$$(\mathrm{A.13}) \qquad \sum q_n = 1 \quad \text{and} \quad q_n < \mathrm{Const}\, \theta^n.$$

Furthermore, the sequence $\{q_n\}$ is independent of ℓ, i.e. it is the same for all standard pairs $\ell = (\gamma, \rho)$ that fully cross Λ^s.

Proof. Due to (A.11), it is easy to define s_ℓ so that that for all $n > 0$

$$(\mathrm{A.14}) \qquad \mathrm{mes}_\ell(x \in \gamma \setminus \gamma_\kappa \colon s_\ell(x) > n)/\mathrm{mes}_\ell(\gamma \setminus \gamma_\kappa) \leq \mathrm{Const}\, \theta^n.$$

We still have a considerable flexibility in defining s_ℓ, and we want to adjust it so that it will satisfy (A.12) with a sequence $\{q_n\}$ independent of ℓ. This seems to be a rigid requirement, but it can be fulfilled by splitting gaps V into 'thinner' curves with the help of rectangles $V \times [0,1]$ described in Step 2: precisely, we can replace each gap V with a rectangle $V \times [0,1]$, divide the latter into subrectangles $V \times I_j$, where $I_j \subset [0,1]$ are some subintervals, and define s_ℓ differently on each subrectangle I_j. The sizes of the subintervals $I_j \subset [0,1]$ must be selected to ensure (A.12), as well as (A.13). $\square$

Step 4. We now turn to the construction of the coupling map $\xi \colon Y_1 \to Y_2$ for Lemma A.3, which will be done recurrently. Given two rectangles Y_1, Y_2 with long bases W_1, W_2, we define the first stopping time to be constant $s_0(x) = n_0$ on both rectangles. At the time $s_0 = n_0$ some of the H-components of each curve W_i will fully cross Λ^s. For every

H-component $W_{1,s_0,i}$ of $\mathcal{F}^{s_0}(W_1)$ that fully crosses Λ^s we consider the corresponding rectangle $Y_{1,s_0,i} = W_{1,s_0,i} \times [0,1]$. We will split off a subrectangle $W_{1,s_0,i} \times [0, \tau_{1,i}]$ with some $\tau_{1,i} \leq 1/2$ so that $m_1(\tilde{Y}_{1,1}) = d_0/2$, where

$$\tilde{Y}_{1,1} = \{(x,t) \in Y_1 \colon \mathcal{F}^{s_0}(x) \in W_{1,s_0,i} \cap \Lambda^s \ \& \ t \in [0, \tau_{1,i}] \text{ for some } i\}$$

(this is possible due to (A.10)).

Suppose we define, similarly, the set $\tilde{Y}_{2,1} \subset Y_2$. Then the sets $\tilde{Y}_{1,1}$ and $\tilde{Y}_{2,1}$ will have the same overall measure ($= d_0/2$), and their $\mathcal{F}^{s_0}$-images will intersect the same stable manifolds $W^s \in \mathcal{G}^s$, but for every $W^s \in \mathcal{G}^s$ the intersections $W^s \cap \mathcal{F}^{s_0}(\tilde{Y}_{1,1})$ and $W^s \cap \mathcal{F}^{s_0}(\tilde{Y}_{2,1})$ may carry different 'amount' of measures m_1 and m_2, respectively. This happens for two reasons: (i) the densities of our measures may vary along H-components and (ii) the Jacobian of the holonomy map may also vary and differ from one. To deal with these problems, we need to assume that the diameter of Λ^s is small, so that the corresponding oscillations of the densities are small (say, the ratio of the densities at different points on the same H-component is between 0.99 and 1.01), and the Jacobian takes values in a narrow interval, say, $[0.99, 1.01]$.

Now we define the set $\tilde{Y}_{2,1}$ as follows. For every H-component $W_{2,s_0,j} \subset \mathcal{F}^{s_0}(W_2)$ that fully crosses Λ^s we will construct a function $\tau_{2,j}(y) \leq 0.6$ on $W_{2,s_0,j} \cap \Lambda^s$ and then put

$$\tilde{Y}_{2,1} = \{(y,t) \in Y_2 \colon \mathcal{F}^{s_0}(y) \in W_{2,s_0,j} \cap \Lambda^s$$
$$\& \ t \in [0, \tau_{2,j}(\mathcal{F}^{s_0}y)] \text{ for some } j\}$$

The functions $\tau_{2,j}$ can be constructed so that for every $W^s \in \mathcal{G}^s$ the intersections $W^s \cap \mathcal{F}^{s_0}(\tilde{Y}_{1,1})$ and $W^s \cap \mathcal{F}^{s_0}(\tilde{Y}_{2,1})$ carry the same 'amount' of measures m_1 and m_2 (this is why we allow $\tau_{2,j}$ to take values up to 0.6). Now we naturally define the coupling map $\xi \colon \tilde{Y}_{1,1} \to \tilde{Y}_{2,1}$ that preserves measures and couples points whose $\mathcal{F}^{s_0}$-images lie on the same stable manifold of the $\mathcal{G}^s$ family. Note that

(A.15) $\qquad m_r(\tilde{Y}_{r,1}) = d_0/2 \qquad \text{for } r = 1,2.$

Lastly we set $R(x,t) = s_0$ on $\tilde{Y}_{1,1}$. This concludes the first round of our recurrent construction of ξ.

Step 5. Before we start the second round, we need to 'inventory' the remaining parts of Y_r, $r = 1, 2$, and represent each of them as a countable union of rectangles. To this end we define a function $\tau_{r,i}$ on every H-component $W_{r,s_0,i}$ of $\mathcal{F}^{s_0}(W_r)$ that fully crosses Λ^s: for $r = 1$ we set $\tau_{1,i}(x)$ to be constant equal to $\tau_{1,i}$ defined in Step 4, and for $r = 2$ we extend the function $\tau_{2,i}(x)$ defined in Step 4 on $W_{2,s_0,i} \cap \Lambda^s$

continuously and linearly to every gap $V_{2,s_0,i,j} \subset W_{2,s_0,i} \setminus \Lambda^s$. The graph of $\boldsymbol{\tau}_{r,i}$ divides the rectangle $W_{r,s_0,i} \times [0,1]$ into two parts ('subrectangles' whose one side may be curvilinear).

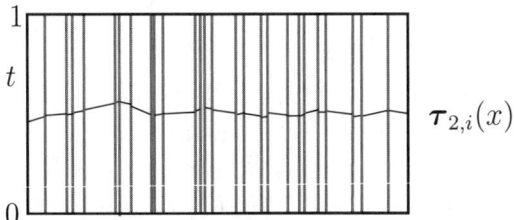

FIGURE 1. The partition of a rectangle over an H-component $W_{2,s_0,i}$: the irregular line in the middle is the graph of the function $\boldsymbol{\tau}_{2,i}(x)$; it separates the 'upper subrectangle' (of the second type) from lower trapezoids (of the third type).

Now the set $\mathcal{F}^{s_0}(Y_r \setminus \tilde{Y}_{r,1})$ consists of connected components of three types. First, there are rectangles corresponding to the H-components of $\mathcal{F}^{s_0}(W_r)$ that do not fully cross Λ^s. Second, the 'upper subrectangles'

$$\{(x,t): x \in W_{r,s_0,i} \ \& \ t \in [\boldsymbol{\tau}_{r,i}(x), 1]\}.$$

These are genuine rectangles for $r = 1$ and figures with one 'jagged' side for $r = 2$, see Fig. 1. All of them have sufficiently long bases (longer than the size of Λ^s in the unstable direction). Third, the 'lower subrectangles'

$$\{(x,t): x \in V_{r,s_0,i,j} \ \& \ t \in [0, \boldsymbol{\tau}_{r,i}(x)]\}.$$

constructed over gaps $V_{r,s_0,i,j} \subset W_{r,s_0,i} \setminus \Lambda^s$. These are true rectangles for $r = 1$ and trapezoids for $r = 2$, see Fig. 1.

The shape of the functions $\boldsymbol{\tau}_{2,i}$ is determined by the densities on our H-components, which are Lipschitz continuous, see (4.20), and the Jacobian of the holonomy map, which is only weakly regular in the following sense. For any two nearby H-curves W', W'' and $x, y \in W'$ that belong to one connected component of $\Omega_\mathcal{D} \setminus \mathcal{S}_n$, the Jacobian of the holonomy map $h: W' \to W''$ satisfies

$$|\ln \mathcal{J}_{W'} h(x) - \ln \mathcal{J}_{W'} h(y)| \leq \mathrm{Const}\, \theta^n$$

for some $\theta < 1$, see [**11**, Theorem 3.6] (this property is sometimes called 'dynamically defined Hölder continuity' [**96**, p. 597]). Thus, the function $\tau_{2,i}$ will be dynamically Hölder continuous, i.e. it will satisfy

(A.16)
$$|\ln \tau_{2,i}(x) - \ln \tau_{2,i}(y)| \leq C\theta^n$$

whenever x and y belong to the same connected component of $\Omega_\mathcal{D} \setminus \mathcal{S}_n$.

Next we rectify the rectangles of the second and third type as follows. Given a 'rectangle' $Y = \{(x,t) \colon x \in W\ \&\ t \in [0, \tau(x)]\}$, where $\tau(x)$ is a dynamically Hölder continuous function on W, equipped with a probability measure $dm(x,t) = \rho(x)\, dx\, dt$, we transform $[0, \tau(x)]$ onto $[0, 1]$ linearly at every point $x \in W$, and thus obtain a full-height rectangle $\hat{Y} = W \times [0, 1]$ with measure

$$d\hat{m}(x,t) = \hat{\rho}(t)\, dx\, dt, \qquad \hat{\rho}(x) = \tau(x)\rho(x).$$

Since $\hat{\rho}(x)$ is dynamically Hölder continuous (rather than Lipschitz), we have to generalize our notion of standard pairs (for the proof of Lemma A.3 only!) to include such densities. This will not do any harm, though, since it is only oscillations of these densities that matters in our proof, and the oscillations are well controlled by the dynamical Hölder continuity; observe also that our densities will smooth out before the next stopping time, thus they will always remain uniformly Hölder continuous – with the same C and θ in (A.16); this is intuitively clear, but see [**20**, page 1089] for the exact argument.

Thus the remaining set $Y_{r,1} \colon= \mathcal{F}^{s_0}(Y_r \setminus \tilde{Y}_{r,1})$ for $r = 1, 2$ is a (countable) union of rectangles of the full (unit) height, which we denote by $\{Y_{r,1,i}\}$; it carries a probability measure $m_{r,1}$ induced by the $\mathcal{F}^{s_0}$-image of the measure m_r. The family $\{Y_{r,1,i}\}$ may not be proper, i.e. it may fail to satisfy (A.6). However, if we condition the measure $m_{r,1}$ onto the union of rectangles of the first and second type, it will obviously recover and become a proper family in just a few iterations of $\mathcal{F}$. On the rectangles of the third type, the recovery time may vary greatly, see Step 3, and we define the stopping time function $s_1(x,t)$ on the rectangles of the third type as described in Lemma A.4. We can clearly define the stopping time s_1 on the rectangles of the first and the second types as well, so that its overall distribution matches that described in Step 3, i.e.

(A.17)
$$m_{r,1}\bigl(Y_{r,1}^{(n)}\bigr) = q_n, \qquad Y_{r,1}^{(n)} \colon= \{\cup_i Y_{r,1,i} \colon s_1 = n\}$$

with the same sequence $\{q_n\}$ as in (A.12)–(A.13). Of course, s_1 must be constant on every rectangle, so to ensure (A.17) we may need to split rectangles $Y_{r,1,i}$ of the first and second type into 'thinner' subrectangles, as we did in the end of Step 3, and define s_1 separately on every subrectangle.

Now for every rectangle $Y_{r,1,i}$ the set $\mathcal{F}^{s_1}(Y_{r,1,i})$ will contain H-components fully crossing Λ^s, and in the notation of (A.9) we have $m_{r,1}(Y_{r,1,i,s_1,*})/m_{r,1}(Y_{r,1,i}) \geq d_0$, due to (A.10), hence

(A.18) $$m_{r,1}(Y^{(n)}_{r,1,s_1,*}) \geq d_0\, m_{r,1}(Y^{(n)}_{r,1}) = d_0 q_n,$$

where $Y^{(n)}_{r,1,s_1,*} = \cup_i Y_{r,1,i,s_1,*}$.

Based on (A.17) and (A.18), for every $n \geq 1$ we can apply our coupling procedure (Step 4) to the sets $Y^{(n)}_{1,1}$ and $Y^{(n)}_{2,1}$ and define the coupling map ξ on a subset of relative measure $d_0/2$, see (A.15), i.e.

(A.19) $$m_{r,1}\bigl((x,t) \in Y^{(n)}_{r,1} \,\&\, \mathcal{F}^n(x,t) \text{ is coupled}\bigr) = d_0 q_n/2$$

We denote by $\tilde{Y}_{r,2} \subset Y_r$ the set of the preimages of just 'coupled' points and put $R(x) = s_0(x) + s_1(\mathcal{F}^{s_0}x)$ on $\tilde{Y}_{1,2}$. We do this for every $n \geq 1$, and this concludes the second round of our construction. It then proceeds recursively, by repeating Steps 4 and 5 alternatively.

At the kth round, we define a stopping time function s_{k-1} on the set $Y_{r,k-1}$ of yet uncoupled points for $r = 1, 2$, then we 'couple' some points of the images $\mathcal{F}^{s_{k-1}}(Y_{r,k-1})$, denote by $\tilde{Y}_{r,k} \subset Y_r$ the set of preimages of just 'coupled' points, and define

$$R(x) = s_0(x) + \cdots + s_{k-1}(\mathcal{F}^{s_0+\cdots+s_{k-2}}x)$$

on $\tilde{Y}_{1,k}$. Observe that the point $\mathcal{F}^{R(x)}(x)$ and its partner $\mathcal{F}^{R(x)}(\xi(x))$ lie on the same stable manifold, which proves the claim (A) of Lemma A.3.

Step 6. It remains to prove the claim (B), which will also imply that the coupling map ξ is defined almost everywhere on Y_1. For brevity, we identify the set Y_r (for each $r = 1, 2$) with its images, i.e. we consider all our stopping time functions as defined on Y_r. We then have two conditional probability formulas:

(A.20) $$m_r(s_k = n | s_{k-1} = n_{k-1}, \ldots, s_1 = n_1, s_0 = n_0) = q_n$$

due to (A.17) and

(A.21) $$m_r\bigl(\tilde{Y}_{r,k} | s_{k-1} = n_{k-1}, \ldots, s_1 = n_1, s_0 = n_0\bigr) = \delta := d_0/2$$

due to (A.19). The following argument is standard in the studies of random walks. Let $\bar{p}_n = m_1\bigl((x,t) \in Y_1 : R(x,t) = n\bigr)$ denote the fraction of points coupled exactly at time n (i.e., at the nth iteration of $\mathcal{F}$, rather than at the nth round). Note that $\bar{p}_i = 0$ for $i < n_0$ and $\bar{p}_{n_0} = \delta$. Now $p_n = \bar{p}_n/\delta$ is the fraction of points *stopped* at time n, i.e.

$$p_n = m_1\bigl((x,t) \in Y_1 : s_0 + s_1 + \cdots + s_k = n \text{ for some } k\bigr).$$

Due to (A.20) and (A.21) we have the following 'convolution law':

$$(A.22) \qquad p_{n+n_0} = (1-\delta)\Big(q_n + (1-\delta)\sum_{i=1}^{n-1} q_{n-i}p_{n_0+i}\Big) \qquad \forall n \geq 1.$$

Now consider two complex analytic functions

$$P(z) = \sum_{n=1}^{\infty} p_{n_0+n}z^n \quad \text{and} \quad Q(z) = \sum_{n=1}^{\infty} q_n z^n,$$

then (A.22) implies $P(z) = (1-\delta)Q(z) + (1-\delta)^2 P(z)Q(z)$, hence

$$(A.23) \qquad P(z) = \frac{(1-\delta)Q(z)}{1-(1-\delta)^2 Q(z)}$$

Due to (A.13), we have $|Q(z)| \leq 1$ for all $|z| \leq 1$, and the function $Q(z)$ is analytic in the complex disk $\{z\colon |z| < 1+\varepsilon\}$ for some $\varepsilon > 0$. Hence $P(z)$ is also analytic in a complex disk of radius greater than one, which implies an exponential tail bound on p_n. A similar bound then follows for $\bar{p}_n = \delta p_n$. $\square$

A.2. Decay of correlations: extensions. In this section we will extend the mixing results in several ways:

Extension 1. Suppose one scatterer (specifically, $\mathbb{B}_0$) is removed from the construction of $\Omega_\mathcal{D}$, i.e. we redefine $\tilde{\Omega}_\mathcal{D} = \cup_{i=1}^{r} \partial \mathbb{B}_i \times [-\pi/2, \pi/2]$, and respectively the return map $\tilde{\mathcal{F}}_\mathcal{D}\colon \tilde{\Omega}_\mathcal{D} \to \tilde{\Omega}_\mathcal{D}$ and the invariant measure $\tilde{\mu}_\mathcal{D}$. Note that we do not change the dynamics – the billiard particle still collides with the scatterer $\mathbb{B}_0$, we simply skip those collisions in the construction of the collision map. Assume, additionally, that the billiard particle cannot experience two successive collisions with $\mathbb{B}_0$ without colliding with some other scatterer(s) in between (this follows from our finite horizon assumption in Section 1.2, provided **r** is small enough). In this case the analysis done in Section A.1 (as well as the earlier one [96]) goes through and Propositions A.2 and A.1 now hold for the dynamical system $(\tilde{\Omega}_\mathcal{D}, \tilde{\mathcal{F}}_\mathcal{D}, \tilde{\mu}_\mathcal{D})$ and functions f, g defined on $\tilde{\Omega}_\mathcal{D}$.

The value of $\theta_{k,\eta}$ in (A.1) depends on the following quantities characterizing the given billiard table:

(a) the minimal and maximal free path (called $L_{\min}$ and $L_{\max}$),
(b) the minimal and maximal curvature of the boundary of the scatterers,
(c) the upper bound on the derivative of the curvature of the scatterers,
(d) the value of $\boldsymbol{\theta}_1$ in the one-step expansion estimate (4.23).

According to (4.24), the value of θ_1 will be bounded away from one because $L_{\max}/L_{\min}$ remains bounded.

We note that the earlier work [96] does not use θ_1. Instead, it uses the *complexity bound*, i.e. the smallest $n \geq 1$ for which

$$K_n \vartheta^n < 1, \tag{A.24}$$

where K_n denotes the maximal number of pieces into which $\mathcal{S}_n$ can partition arbitrary short unstable curves. It is known [10, Section 8] and [96, p. 634] that $K_n \leq C_1 n + C_2$, where C_1 and C_2 are constants determined by the number of possible tangencies between successive collisions, i.e. by the maximal number of points at which a straight line segment $I \subset \mathcal{D}$ can touch some scatterers $\mathbb{B}_i$. We note that this number does not exceed $L_{\max}/L_{\min}$, thus C_1, C_2, and n in (A.24) are effectively determined by $L_{\max}/L_{\min}$ which remains bounded.

Extension 2. Consider a family of dispersing billiard tables obtained by changing the position of one of the scatterers (specifically, $\mathbb{B}_0$) continuously on the original dispersing billiard table. We only allow such changes that the maximal free path $L_{\max}$ remains bounded away from infinity, and the minimal free path $L_{\min}$ remains bounded away from zero. Then all the characteristic values (a)–(d) of the billiard tables in our family will effectively remain unchanged, and therefore the bound (A.1) will be *uniform*. (Note that the space $\Omega_\mathcal{D}$ does not depend on the position of the movable scatterer $\mathbb{B}_0$, hence the functions f, g in (A.1) do not have to change with the position of $\mathbb{B}_0$).

Extension 3. Suppose we not only change the position of the scatterer $\mathbb{B}_0$, but also reduce its size homotetically (namely, suppose $\mathbb{B}_0$ is a disk of radius $\mathbf{r}_0$, and we replace it with a disk of radius $\mathbf{r} < \mathbf{r}_0$). Hence we consider a larger family of dispersing billiard tables than in Extension 2. Now the collision space $\Omega_\mathcal{D}$ depends on the size of $\mathbb{B}_0$, but we restrict the analysis to the space $\tilde{\Omega}_\mathcal{D}$ constructed exactly as we did in Extension 1, by skipping collisions with $\mathbb{B}_0$. Then the space $\tilde{\Omega}_\mathcal{D}$ will be the same for all billiard tables in our family, so we can speak about the uniformity of the exponential bound on correlations for the map $\tilde{\mathcal{F}}$. Again, we assume a uniform upper bound on $L_{\max}$ and a uniform positive lower bound on $L_{\min}$. There are several new problems now:

The curvature of $\partial \mathbb{B}_0$ will not be uniformly bounded anymore, it will be proportional to $1/\mathbf{r}$. The upper bound on the curvature is used to prove a uniform transversality of stable and unstable cones, see [18, pp. 534–535]. Those cones are not uniformly transversal anymore, the angle between them is $\mathcal{O}(\mathbf{r})$ on the part of the phase space $\Omega_\mathcal{D}$ corresponding to the boundary of the scatterer $\mathbb{B}_0$, but this part is specifically excluded from the construction of $\tilde{\Omega}_\mathcal{D}$, hence the cones are still uniformly transversal on $\tilde{\Omega}_\mathcal{D}$. The upper bound on the curvature is also used in the distortion and curvature estimates, similar to those in Appendix C, but we will show that those estimates remain uniform over all $\mathbf{r} > 0$, see a remark after the proof of Lemma C.1. Next, the curvature of the disk $\mathbb{B}_0$ is constant, so its derivative is zero.

Lastly, the complexity K_n of the singularity set $\mathcal{S}_n$ will be affected by $\mathbf{r}$, too, if $\mathbf{r}$ is allowed to be arbitrarily small. Indeed, if all the scatterers had fixed size, one considers [**18, 96**] short enough unstable curves that can only break into two pieces at any collision (one piece collides, the other passes by, as it is explained in the proof of Lemma 4.10). But now, no matter how small an unstable curve is, the scatterer $\mathbb{B}_0$ may be even smaller, and then the unstable curve may be torn by $\mathbb{B}_0$ into *three* pieces. The middle piece hits $\mathbb{B}_0$, gets reflected, and by the next collision its image will be of length $\mathcal{O}(1)$. It is not hard to see then that the sequence K_n will grow exponentially fast, and therefore the complexity bound (A.24) may easily fail for all $n \geq 1$.

The complexity bound is only used in the proof of the growth lemma [**18**, Theorem 3.1], which is analogous to our Lemma 4.10. We have seen in Section 4.4 that the growth lemma follows from the one-step expansion estimate (4.23). In fact, it suffices to establish the one-step expansion estimate for any iteration of the given map, see [**18**, Proposition 10.1] and [**30**, Theorem 10], i.e. in our case it is enough to prove that

$$(A.25) \qquad \exists n \geq 1: \quad \boldsymbol{\theta}_n := \liminf_{\delta \to 0} \sup_{W:\, \text{length}(W) < \delta} \sum_i \vartheta_{i,n} < 1$$

Here $W \subset \tilde{\Omega}_\mathcal{D}$ denotes an H-curve and $\vartheta_{i,n}^{-1}$ the smallest local factor of expansion of $\tilde{\mathcal{F}}_\mathcal{D}^{-n}(W_{i,n})$ under the map $\tilde{\mathcal{F}}_\mathcal{D}^n$, where $W_{i,n}$, $i \geq 1$, denote the H-components of $\tilde{\mathcal{F}}_\mathcal{D}^n(W)$.

Next we prove (A.25). First we consider the case $n = 1$. Collisions of W with the fixed scatterers $\mathbb{B}_j$, $j \geq 1$, are described in the proof of Lemma 4.10. Now if W collides with the disk $\mathbb{B}_0$ of a very small radius $\mathbf{r}$, say $\mathbf{r} < \mathcal{O}(\text{length}(W))$, then W may be torn into three pieces as described above. The middle piece (reflecting off $\mathbb{B}_0$) will be further subdivided into countably many H-components lying in all the homogeneity strips $\mathbb{H}_{\pm k}$ for $k \geq k_0$, as well as $\mathbb{H}_0$. Those H-components will be expanded by factors $\geq ck^2/\mathbf{r}$ and $> c/\mathbf{r}$, respectively, see our estimates in Section 4.1. Hence the contribution of all these H-components to the sum $\sum \vartheta_{i,1}$ will be $\mathbf{r}/c + 2\mathbf{r} \sum_{k \geq k_0} (ck^2)^{-1} \leq \text{Const } \mathbf{r}$.

Thus, the image $\tilde{\mathcal{F}}_{\mathcal{D}}(W)$ may consist, generally, of the following H-components $W_{i,1}$: countably many $W_{i,1}$'s produced by a collision with $\mathbb{B}_0$, at most $L_{\max}/L_{\min}$ countable sets of $W_{i,1}$'s produced by almost tangential reflections off some fixed scatterers (cf. the proof of Lemma 4.10), and at most two H-components that miss the collision with $\mathbb{B}_0$ and all the grazing collisions – these land somewhere else on $\partial \mathcal{D}$. The last two H-components are only guaranteed to expand by a moderate factor of ϑ^{-1}, which gives an estimate

$$\text{(A.26)} \qquad \boldsymbol{\theta}_1 \leq 2\vartheta + \frac{L_{\max}}{L_{\min}}\frac{\text{Const}}{k_0} + \text{Const}\,\mathbf{r}$$

Note that if $\mathbf{r} > \text{length}(W) = \mathcal{O}(\delta)$, then there is at most one (not two) H-component expanding by ϑ^{-1}, and then (A.26) could be easily handled as in the proof of Lemma 4.10 (a). Thus, we may assume that $\mathbf{r} = \mathcal{O}(\delta)$, and taking $\limsup_{\delta \to 0}$ we can simplify (A.26) as

$$\boldsymbol{\theta}_1 \leq 2\vartheta + \text{Const}/k_0$$

The last term can be made arbitrarily small by selecting k_0 large, as in the proof of Lemma 4.10 (a), but the first term may already exceed one, hence the estimate (A.25) would fail for $n = 1$.

Therefore, we have to consider the case $n \geq 2$. Our previous analysis shows that the image $\tilde{\mathcal{F}}_{\mathcal{D}}^n(W)$ will consist of H-components $W_{i,n}$ of two general types: (a) countably many H-components that have either collided with $\mathbb{B}_0$ at least once or got reflected almost tangentially off some fixed scatterer at least once, and (b) all the other H-components. Respectively, we decompose $\sum_i \vartheta_{i,n} = \sum^{(a)} + \sum^{(b)}$.

First, we estimate $\sum^{(a)}$. The above estimate (A.26) can be easily extended to a more general bound:

$$\Theta := \sup_W \sum_i \vartheta_{i,1} < \text{Const}$$

where the supremum is taken over all H-curves $W \subset \tilde{\Omega}_{\mathcal{D}}$. Now the chain rule and the induction on n gives

$$\text{(A.27)} \qquad \sum\nolimits^{(a)} \leq \text{Const}\,\Theta^n(\mathbf{r} + 1/k_0)$$

We now turn to $\sum^{(b)}$. First, we need to estimate the maximal number of H-components of type (b), we call it $\tilde{K}_n$. Suppose for a moment that $\mathbb{B}_0$ is removed from the billiard table. Then any short u-curve will be cut into at most $K'_n \leq C_1 n + C_2$ pieces by the singularities of the corresponding collision map during the first n collisions, see above, where C_1 and C_2 only depend on the fixed scatterers $\mathbb{B}_i$, $i \geq 1$. Now we put $\mathbb{B}_0$ back on the table. As we have seen, each unstable curve during a free flight between successive collisions with the fixed scatterers can be cut by $\mathbb{B}_0$ into three pieces, of which only two (the middle one excluded) can produce H-components of type (b), thus adding one more piece to our count. Therefore the total number of pieces of type (b), after n reflections, will not exceed $\tilde{K}_n \leq nK'_n \leq C_1 n^2 + C_2 n$. This gives a quadratic bound on $\tilde{K}_n$, and it is important that this bound is independent of the location or the size of the variable scatterer $\mathbb{B}_0$, i.e. our bound is uniform over all the billiard tables in our family.

Now, since $\vartheta_{i,n} \leq \vartheta^n$ for every H-component of type (b), then

$$\sum\nolimits^{(b)} \leq \tilde{K}_n \vartheta^n \leq (C_1 n^2 + C_2 n)\, \vartheta^n$$

thus

$$\boldsymbol{\theta}_n \leq (C_1 n^2 + C_2 n)\, \vartheta^n + \mathrm{Const}\, \Theta^n / k_0$$

where the $\limsup_{\delta \to 0}$ is already taken to eliminate $\mathbf{r}$ from (A.27). Clearly the first term here is less than one for some $n \geq 1$, and then the second term can be made arbitrarily small by choosing k_0 large, hence we obtain (A.25).

This proves that the exponential bound on correlations for the map $\tilde{\mathcal{F}}_\mathcal{D}$ will be *uniform* for all the billiard tables in the family constructed in Extension 3.

We need to make yet another remark: the one-step expansion estimate (A.25) implies the analogue of the growth lemma 4.10 for the map $\tilde{\mathcal{F}}$, with all the constants $\beta_1, \ldots, \beta_6$ and q independent of the location or the size of $\mathbb{B}_0$.

A.3. Large deviations. Consider an unstable curve W with the Lebesgue measure $d\nu$ on it. Denote by $\mathcal{J}_W \mathcal{F}^n(x)$ the Jacobian (the expansion factor) of the map $\mathcal{F}^n$ restricted to W at the point $x \in W$.

PROPOSITION A.5 (Large deviations). *There are constants $K > 0$ and $\theta < 1$ such that uniformly in W and $n \geq 1$*

$$\nu\big(x \in W\colon\ \ln \mathcal{J}_W \mathcal{F}^n(x) > Kn\big) \leq \mathrm{Const}\, \theta^n.$$

Note: by time reversibility, a similar estimate holds for stable curves and negative iterations of $\mathcal{F}$.

LEMMA A.6. *There is $A > 1$ such that for any $\zeta \in (0, 1/2)$ there is $C_\zeta > 0$ such that uniformly in W and n*

$$\int_W \big|\mathcal{J}_W \mathcal{F}^n(x)\big|^\zeta d\nu \leq C_\zeta A^n.$$

Proof. For every $x \in W$ let $m_1, \ldots, m_n$ be the indices of the homogeneity strips where the first n images of x belong, i.e. let $\mathcal{F}^i(x) \in \mathbb{H}_{m_i}$ for $1 \leq i \leq n$. To avoid zeroes, let us relabel the set $\mathbb{H}_0$ by $\mathbb{H}_1$ here. Now, as we mentioned in Section A.1, the expansion factor of $\mathcal{F}$ on u-curves $W \subset \mathcal{F}^{-1}(\mathbb{H}_m)$ is $\mathcal{O}(m^2)$, hence

$$C_1^n \, m_1^2 \cdots m_n^2 < \mathcal{J}_W \mathcal{F}^n(x) < C_2^n \, m_1^2 \cdots m_n^2$$

for some constants $C_2 > C_1 > 0$. On the other hand, there is a constant $B > 1$ such that for any given sequence $m_1, \ldots, m_n$ ("itinerary"), there is at most B^n H-components $W_k \subset \mathcal{F}^n(W)$ so that the points $x \in \mathcal{F}^{-n}(W_k)$ have exactly this itinerary. This fact can be proved by induction on n: given an H-component W_k, its image has at most B H-components in every single homogeneous strip $\mathbb{H}_m$, cf. Section A.1, where $B = L_{\max}/L_{\min}$.

Therefore, denoting by $W_{m_1 \ldots m_n}$ the set of points $x \in W$ with the given itinerary $m_1, \ldots, m_n$ we obtain

$$\nu(W_{m_1 \ldots m_n}) < (B/C_1)^n \, m_1^{-2} \cdots m_n^{-2}$$

Hence

$$\int_W |\mathcal{J}_W \mathcal{F}^n(x)|^\zeta \, d\nu \leq \sum_{m_1, \ldots, m_n} (BC_2^\zeta/C_1)^n \, m_1^{-2+2\zeta} \cdots m_n^{-2+2\zeta}$$

and the series converges for any $\zeta < 1/2$. □

Proof of Proposition A.5 is based on Lemma A.6 and Markov inequality:

$$\nu\bigl(x \in W\colon \ln \mathcal{J}_W \mathcal{F}^n(x) > Kn\bigr) = \nu\bigl(x \in W\colon |\mathcal{J}_W \mathcal{F}^n(x)|^\zeta > e^{\zeta Kn}\bigr)$$
$$\leq C_\zeta [A \exp(-\zeta K)]^n.$$

It remains to choose K so large that $A \exp(-\zeta K) < 1$. □

A.4. Moderate deviations. We use the notation of the previous section and denote by χ the positive Lyapunov exponent of the map $\mathcal{F}$.

PROPOSITION A.7 (Moderate deviations). *Given $\delta > 0$, there are constants $C, a > 0$ such that*

$$\nu\bigl(x \in W\colon |\ln \mathcal{J}_W \mathcal{F}^n(x) - n\chi| > k\bigr) \leq C \exp(-ak^2/n)$$

uniformly in W, $n > 0$ and $\sqrt{n} \leq k \leq n^{2/3-\delta}$

Note: by time reversibility, χ is also the positive Lyapunov exponent of the map $\mathcal{F}^{-1}$, and the above estimate holds for stable curves and negative iterations of $\mathcal{F}$.

Proof. We can assume that W is long enough (for example, $|W| \geq \varepsilon_0$, see Section A.1) and replace ν with a smooth probability measure on W; i.e. we replace (W, ν) with a standard pair $\ell = (\gamma, \rho)$. Let W_x^u denote the unstable manifold through $x \in \Omega$. Since the tangent lines $\mathcal{T}_{\mathcal{F}^i x}(\mathcal{F}^i \gamma)$ and $\mathcal{T}_{\mathcal{F}^i x}(W_{\mathcal{F}^i x}^u)$ are getting exponentially close to each other as $i \to \infty$, the difference between $\ln \mathcal{J}_\gamma \mathcal{F}^n(x) = \sum_{i=0}^{n-1} \ln \mathcal{J}_{\mathcal{F}^i \gamma} \mathcal{F}(\mathcal{F}^i x)$ and $\sum_{i=0}^{n-1} \ln \mathcal{J}_{W_{\mathcal{F}^i x}^u} \mathcal{F}(\mathcal{F}^i x)$ is bounded uniformly in n; so it is enough to prove

(A.28) $$\operatorname{mes}_\ell (x \in \gamma \colon |S_n| > k) \leq C \exp(-ak^2/n),$$

where

(A.29) $$S_n = \sum_{i=0}^{n-1} A \circ \mathcal{F}^i, \qquad A(x) = \ln \mathcal{J}_{W_x^u} \mathcal{F}(x) - \chi.$$

Next we pick $m = m(n)$ such that

(A.30) $$k^2/n \ll m \ll n/k \ll m^{100}$$

where $P \ll Q$ means that $P/Q = \mathcal{O}(n^{-\varepsilon})$ for some $\varepsilon > 0$. For example, $m = n^{1/3}$ will suffice.

Next we divide the time interval $[0, n]$ into segments of length m; we will estimate the sums over odd-numbered intervals and those over even-numbered intervals separately[1]. Accordingly, we define

$$R_j^{(1)} := \sum_{i=2jm-m}^{2jm-1} A \circ \mathcal{F}^i, \qquad R_j^{(2)} := \sum_{i=2jm}^{2jm+m-1} A \circ \mathcal{F}^i,$$

for $1 \leq j < L := \frac{n}{2m}$. Then we denote $Z_r^{(1)} = \sum_{j=1}^r R_j^{(1)}$ and $Z_r^{(2)} = \sum_{j=1}^r R_j^{(2)}$ for $r \leq L$, and obtain $S_n = S_m + Z_L^{(1)} + Z_L^{(2)}$, so

(A.31) $$\operatorname{mes}_\ell(|S_n| > k) \leq \operatorname{mes}_\ell(|S_m| > k/3) + \sum_{j=1,2} \operatorname{mes}_\ell(|Z_L^{(j)}| > k/3).$$

The first term with S_m will be handled later. Our analysis of $Z_L^{(1)}$ and $Z_L^{(2)}$ is completely similar, so we will do it for $Z_L^{(1)}$ only (and omit the superscript (1) for brevity).

---

[1]Thus our method resembles the big-small block technique of probability theory, except our blocks have the same length. It seems that using blocks of variable lengths may help to optimize the value of a in Proposition A.7, but we do not pursue this goal.

LEMMA A.8. *There exists a subset $\hat{\gamma} \subset \gamma$ such that*

(A.32) $$\mathrm{mes}_\ell(\hat{\gamma}) \leq \mathrm{Const}\, e^{-k^2/n}$$

and for every $m^{-100} < |t| < m^{-1}$

(A.33) $$\int_{\gamma \setminus \hat{\gamma}} e^{tZ_L}\, d\mathrm{mes}_\ell \leq e^{Dt^2 n}$$

where $D > 1$ is a constant.

This lemma implies
$$\mathrm{mes}_\ell(Z_L > k) \leq \mathrm{mes}_\ell(\hat{\gamma}) + e^{Dt^2 n - tk}.$$
Substitution $t = \frac{k}{2Dn}$ (which is between m^{-100} and m^{-1} due to (A.30)) gives $\mathrm{mes}_\ell(Z_L > k) \leq \mathrm{Const}\, e^{-ak^2/n}$ with $a = 1/4D$. Similarly we obtain $\mathrm{mes}_\ell(Z_L < -k) \leq \mathrm{Const}\, e^{-ak^2/n}$, and combining we get

(A.34) $$\mathrm{mes}_\ell(|Z_L| > k) \leq \mathrm{Const}\, e^{-ak^2/n},$$

which takes care of $Z_L = Z_L^{(1)}$ in (A.31).

Proof of Lemma A.8. We construct, inductively, sets $\emptyset = \hat{\gamma}_0 \subset \hat{\gamma}_1 \subset \cdots \subset \hat{\gamma}_L =: \hat{\gamma}$ such that (i) the image $\mathcal{F}^{2mr}(\hat{\gamma}_r)$ is a union of some H-components of the set $\mathcal{F}^{2mr}(\gamma)$, (ii) $\mathrm{mes}_\ell(\hat{\gamma}_r \setminus \hat{\gamma}_{r-1}) \leq \mathrm{Const}\, \theta^m$ for some constant $\theta < 1$, and (iii) we have

(A.35) $$\int_{\gamma \setminus \hat{\gamma}_r} e^{tZ_r}\, d\mathrm{mes}_\ell \leq e^{Dt^2 mr}.$$

Then (ii) implies (A.32), since $m(\hat{\gamma}) = \mathcal{O}(L\theta^m) = \mathcal{O}(e^{-k^2/n})$ due to (A.30).

Suppose $\hat{\gamma}_r$ is constructed. Let $\gamma_{r,\alpha}$ denote all the H-components of the set $\mathcal{F}^{2mr}(\gamma \setminus \hat{\gamma}_r)$ and $c > 0$ a small constant. We put

(A.36) $$\gamma_r^{(c)} = \cup_\alpha \{\gamma_{r,\alpha} \colon |\gamma_{r,\alpha}| < e^{-cm}\}, \qquad \hat{\gamma}_r^{(1)} = \mathcal{F}^{-2mr}(\gamma_r^{(c)}).$$

By Lemma 4.10 (b), $\mathrm{mes}_\ell(\hat{\gamma}_r^{(1)}) = \mathcal{O}(e^{-cm})$.

Next let $\gamma_{r,\alpha} \not\subseteq \gamma_r^{(c)}$ be one of the 'longer' components, denote by $\tilde{\rho}_{r,\alpha}$ the induced density on $\gamma_{r,\alpha}$ and put

(A.37) $$\rho_{r,\alpha,t} = \frac{\tilde{\rho}_{r,\alpha}\, e^{tZ_r \circ \mathcal{F}^{-2mr}}}{\int_{\gamma_{r,\alpha}} \tilde{\rho}_{r,\alpha}\, e^{tZ_r \circ \mathcal{F}^{-2mr}}\, dx}.$$

The function $A(x)$ defined by (A.29) is smooth along unstable manifolds, hence $\ell_{r,\alpha,t} = (\gamma_{r,\alpha}, \rho_{r,\alpha,t})$ is a standard pair, and the regularity of $\rho_{r,\alpha,t}$ is uniform in r, α, and $|t| < 1/m$. Even though $A(x)$ is not smooth over Ω, it is 'dynamically Hölder continuous' in the sense of (A.16), see [11, Theorem 3.6]. Now the same argument as in the proof of Proposition A.2, which is based on Lemma A.3, implies $|\mathbb{E}_{\ell_{r,\alpha,t}}(A \circ \mathcal{F}^i)| \leq \mathrm{Const}\,\theta^i$ for some $\theta < 1$ and all $i \geq m$, provided c in (A.36) is small enough, namely we need $(1-c)/c > K$, where K is the constant from Proposition A.2 (observe that $\int_\Omega A\,d\mu = 0$). Hence we have

(A.38) $\quad |\mathbb{E}_{\ell_{r,\alpha,t}}(\tilde{R}_{r+1})| \leq \mathrm{Const}\,\theta^m, \quad |\mathbb{E}_{\ell_{r,\alpha,t}}(\tilde{R}_{r+1}^2)| \leq \mathrm{Const}\,m$

where $\tilde{R}_{r+1} = R_{r+1} \circ \mathcal{F}^{-2mr}$; the second bound follows by the same argument as in Chapter 6.

Next let $\tilde{\gamma}_\beta$ denote all the H-components of $\mathcal{F}^{2m(r+1)}(\gamma \setminus \hat{\gamma}_r)$ and

$$\gamma_r^{(K)} := \cup_\beta \{\mathcal{F}^{-2m}(\tilde{\gamma}_\beta) : \max_{x \in \mathcal{F}^{-2m}(\tilde{\gamma}_\beta)} |\tilde{R}_{r+1}(x)| \geq Km\},$$

where $K > 0$ is the constant from the Proposition A.5 on large deviations. Put $\hat{\gamma}^{(2)} = \mathcal{F}^{-2mr}(\gamma_r^{(K)})$. Since the oscillations of $\tilde{R}_{r+1}$ on each curve $\mathcal{F}^{-2m}(\tilde{\gamma}_\beta)$ are $\mathcal{O}(1)$, it easily follows from Proposition A.5 that $\mathrm{mes}_\ell(\hat{\gamma}_r^{(2)}) = \mathcal{O}(\theta^m)$. Now the set $\hat{\gamma}_{r+1} := \hat{\gamma}_r \cup \hat{\gamma}_r^{(1)} \cup \hat{\gamma}_r^{(2)}$ will satisfy the requirements (i) and (ii), so it remains to prove (iii).

Let $\gamma_{r,\alpha} \not\subseteq \gamma_r^{(c)}$. For brevity, denote $\gamma' = \gamma_{r,\alpha} \setminus \gamma_r^{(K)}$ and $\gamma'' = \gamma_{r,\alpha} \cap \gamma_r^{(K)}$, as well as $\rho = \rho_{r,\alpha,t}$. At every point $x \in \gamma'$ we have $|\tilde{R}_{r+1}| < Km$, hence

$$e^{t\tilde{R}_{r+1}} \leq 1 + t\tilde{R}_{r+1} + At^2 \tilde{R}_{r+1}^2$$

with a constant $A > 1$, uniformly in $|t| < 1/m$. Thus

$$\int_{\gamma'} e^{t\tilde{R}_{r+1}} \rho\,dx \leq \int_{\gamma'} (1 + t\tilde{R}_{r+1} + At^2 \tilde{R}_{r+1}^2)\rho\,dx$$

$$\leq \int_{\gamma' \cup \gamma''} (1 + t\tilde{R}_{r+1} + At^2 \tilde{R}_{r+1}^2)\rho\,dx$$

$$\leq 1 + Bt^2 m \leq e^{Bt^2 m}$$

with some constant $B > 0$. To obtain the second line, we used

$$\int_{\gamma''} |t\tilde{R}_{r+1}|\rho\,dx \leq \int_{\gamma''} (1 + t^2 \tilde{R}_{r+1}^2)\rho\,dx,$$

and for the third line we used (A.38) (note that $|t|\theta^m \ll t^2 m$ since $t \geq m^{-100}$). Now using (A.37) gives

$$\int_{\gamma'} \tilde{\rho}_{r,\alpha} e^{tZ_{r+1} \circ \mathcal{F}^{-2mr}}\,dx = \int_{\gamma'} e^{t\tilde{R}_{r+1}} \rho\,dx \times \int_{\gamma_{r,\alpha}} \tilde{\rho}_{r,\alpha} e^{tZ_r \circ \mathcal{F}^{-2mr}}\,dx$$

$$\leq e^{Bt^2 m} \int_{\gamma_{r,\alpha}} \tilde{\rho}_{r,\alpha} e^{tZ_r \circ \mathcal{F}^{-2mr}}\,dx$$

Summation over α and using (A.35) implies

$$\int_{\gamma\setminus\hat{\gamma}_{r+1}} e^{tZ_{r+1}}\,\mathrm{dmes}_\ell \leq e^{Dt^2 m(r+1)}$$

(provided $D \geq B$), which proves (A.35) inductively. $\square$

It remains to handle the first term in (A.31). Note that $k \gg m$, and due to the uniform hyperbolicity $A \geq 0$, hence $S_m \geq -\chi m$. The necessary upper bound on S_m will follow from the next lemma, which is similar to Proposition A.5 on large deviations, but it controls "very large deviations":

LEMMA A.9. *We have* $\mathrm{mes}_\ell(S_m > k) \leq \mathrm{Const}\, m\, e^{-k/m}$ *for all* $k > 0$.

Proof. If $S_m(x) > k$, then $A \circ \mathcal{F}^i(x) > k/m$ for some $0 \leq i < m$, therefore $\mathcal{F}^{i+1}(x)$ lies in the $(e^{-k/m})$-neighborhood of $\mathcal{S}_0 = \partial\Omega$, but for each i the probability of this event is $\leq \mathrm{Const}\, e^{-k/m}$ due to the growth lemma 4.10. This completes the proof of the lemma and that of Proposition A.7. $\square$

A.5. Nonsingularity of diffusion matrix. Here we discuss the properties of the matrix $\bar{\sigma}_Q^2(\mathcal{A})$ defined by the Green-Kubo formula (1.14).

LEMMA A.10. *The matrix* $\bar{\sigma}_Q^2(\mathcal{A})$ *depends on* Q *continuously.*

Proof. Every term in the series (1.14) depends on Q continuously, and the claim now follows from a uniform bound proved in Extension 1 of Section A.1. $\square$

Next we describe the conditions under which the matrix $\bar{\sigma}_Q^2(\mathcal{A})$ is nonsingular. For any vector $u \in \mathbb{R}^2$ we have

$$u^T \bar{\sigma}_Q^2(\mathcal{A})\, u = \sum_{n=-\infty}^{\infty} \int_{\Omega_Q} (u^T \mathcal{A})\big[(\mathcal{A} \circ \mathcal{F}_Q^n)^T u\big]\,d\mu_Q$$

$$= \sum_{n=-\infty}^{\infty} \int_{\Omega_Q} g_u\,(g_u \circ \mathcal{F}_Q^n)\,d\mu_Q$$

where $g_u = \langle \mathcal{A}, u \rangle$ is a smooth function on Ω_Q.

Fact. For any smooth function $g \colon \Omega_Q \to \mathbb{R}$, the following three conditions are equivalent:

1. $\sum_{n=-\infty}^{\infty} \int g_u\,(g_u \circ \mathcal{F}_Q^n)\,d\mu_Q = 0$;
2. $g = h \circ \mathcal{F}_Q - h$ for some $h \in L^2(\Omega_Q)$;

3. For any periodic point $x \in \Omega_Q$ with period $k \geq 1$, such that g is smooth at $x, \mathcal{F}(x), \ldots, \mathcal{F}^{k-1}(x)$ we have

$$S_g(x) := \sum_{i=0}^{k-1} g(\mathcal{F}_Q^i x) = 0$$

The equivalence of 1 and 2 is a standard fact of ergodic theory (see e.g. [32, Lemma 2.2]); for the equivalence of 2 and 3 see [38], [11, Section 7], and [13, Section 5].

Now, if the matrix $\bar{\sigma}_Q^2(\mathcal{A})$ is singular, it has an eigenvector u corresponding to the zero eigenvalue, so that $\bar{\sigma}_Q^2(\mathcal{A}) = 0$. Equivalently, for any periodic point $x \in \Omega_Q$ of period $k \geq 1$,

$$\langle S_\mathcal{A}(x), u \rangle = 0, \qquad S_\mathcal{A}(x) := \sum_{i=0}^{k-1} \mathcal{A}(\mathcal{F}_Q^i x) = 0$$

Therefore, we obtain the following:

Criterion for nonsingularity of $\bar{\sigma}_Q^2(\mathcal{A})$. Suppose there are two periodic points, $x_1, x_2 \in \Omega$ with periods k_1 and k_2, respectively, such that the vectors $S_\mathcal{A}(x_1)$ and $S_\mathcal{A}(x_2)$ are nonzero and noncollinear. Then $\bar{\sigma}_Q^2(\mathcal{A})$ is nonsingular.

Observe that there is at least one periodic point x such that $S_\mathcal{A}(x) \neq 0$, it is made by an orbit running between $\mathcal{P}(Q)$ and the closest scatterer $\mathbb{B}_i$. On many billiard tables, one can easily find several such trajectories, which would guarantee the nondegeneracy of $\bar{\sigma}_Q^2(\mathcal{A})$.

A.6. Asymptotics of diffusion matrix. Here we discuss the asymptotics of $\sigma_Q^2(\mathcal{A})$ as $\mathbf{r} \to 0$ and prove (2.20), in fact a stronger version of it:

$$\sigma_Q^2(\mathcal{A}) = \frac{8\mathbf{r}}{3\,\text{Area}(\mathcal{D})} I + Z_Q \mathbf{r}^2 + o(\mathbf{r}^2),$$

where Z_Q is a 2×2 matrix (independent of $\mathbf{r}$). By virtue of (1.17), this is equivalent to

$$(\text{A.39}) \quad \bar{\sigma}_Q^2(\mathcal{A}) = \frac{8\pi\mathbf{r}}{3\,\text{length}(\partial\mathcal{D})} I + \left(Z_Q - \frac{16\pi^2}{3\,[\text{length}(\partial\mathcal{D})]^2} I\right) \mathbf{r}^2 + o(\mathbf{r}^2).$$

We also provide an explicit algorithm for computing Z_Q.

First we fix our notation. To emphasize the dependence of our dynamics on $\mathbf{r}$ we denote by $\Omega_{Q,\mathbf{r}}$ the collision space, $\mathcal{F}_{Q,\mathbf{r}}$ the collision map and $\mu_{Q,\mathbf{r}}$ the invariant measure. We also use notation of Extension 3 of Section A.1, after identifying our disk $\mathcal{P}(Q)$ with the variable scatterer $\mathbb{B}_0$: thus we get the collision space $\tilde{\Omega}_\mathcal{D}$, the collision map $\tilde{\mathcal{F}}_\mathcal{D}$

on it (however, we will denote this map by $\tilde{\mathcal{F}}_{Q,\mathbf{r}}$ to emphasize its dependence on Q and $\mathbf{r}$), and the corresponding invariant measure $\tilde{\mu}_\mathcal{D}$. Note that $\tilde{\mu}_\mathcal{D}$ is obtained by conditioning the measure $\mu_{Q,\mathbf{r}}$ on $\tilde{\Omega}_\mathcal{D}$, the ratio of their densities is

$$(\text{A.40}) \qquad L_\mathbf{r} := \frac{\text{length}(\partial \mathcal{D}) + 2\pi \mathbf{r}}{\text{length}(\partial \mathcal{D})},$$

and $\tilde{\mu}_\mathcal{D}$ is in fact independent of Q and $\mathbf{r}$.

Consider the function $\tilde{\mathcal{A}}(x) := \mathcal{A}(\mathcal{F}_Q(x))$ on $\tilde{\Omega}_\mathcal{D}$ and the matrix

$$(\text{A.41}) \qquad \tilde{\sigma}_Q^2(\tilde{\mathcal{A}}) := \sum_{n=-\infty}^{\infty} \int_{\tilde{\Omega}_\mathcal{D}} \tilde{\mathcal{A}} \left(\tilde{\mathcal{A}} \circ \tilde{\mathcal{F}}_{Q,\mathbf{r}}^n \right)^T d\tilde{\mu}_\mathcal{D}.$$

It follows from [69, Theorem 1.3] that $\tilde{\sigma}_Q^2(\tilde{\mathcal{A}}) = L_\mathbf{r} \bar{\sigma}_Q^2(\mathcal{A})$. Hence it is enough to prove that

$$(\text{A.42}) \qquad \tilde{\sigma}_Q^2(\tilde{\mathcal{A}}) = \frac{8\pi \mathbf{r}}{3\,\text{length}(\partial \mathcal{D})} I + Z_Q \mathbf{r}^2 + o(\mathbf{r}^2).$$

First we will establish a weaker formula

$$(\text{A.43}) \qquad \tilde{\sigma}_Q^2(\tilde{\mathcal{A}}) = \frac{8\pi \mathbf{r}}{3\,\text{length}(\partial \mathcal{D})} I + \mathcal{O}(\mathbf{r}^2 \ln \mathbf{r}),$$

which is, by the way, sufficient for our main purpose of proving (2.20), and then outline a proof of the sharp estimate (A.39).

Due to the invariance of the measure $\mu_{Q,\mathbf{r}}$ under the map $\mathcal{F}_{Q,\mathbf{r}}$, we have $\tilde{\mu}_\mathcal{D}(\tilde{\mathcal{A}}) = \mu_{Q,\mathbf{r}}(\mathcal{A}) = 0$. It is easy to check that $\tilde{\mathcal{A}}$ is Hölder continuous with exponent $\eta = 1/2$ and coefficient $K_{\tilde{\mathcal{A}}} = \text{Const}/\mathbf{r}$, hence $\tilde{\mathcal{A}} \in \mathcal{H}_{1,1/2}$, in the notation of Section A.1. Therefore, the uniform bound on correlations proved in Extension 3 gives

$$(\text{A.44}) \qquad \left| \int_{\tilde{\Omega}_\mathcal{D}} \tilde{\mathcal{A}} \left(\tilde{\mathcal{A}} \circ \tilde{\mathcal{F}}_{Q,\mathbf{r}}^n \right)^T d\tilde{\mu}_\mathcal{D} \right| \leq \text{Const}\, \mathbf{r}^{-2} \theta_{1,1/2}^{|n|}$$

where Const is independent of Q and $\mathbf{r}$. Let K be such that $\theta_{1,1/2}^{K|\ln \mathbf{r}|} = \mathbf{r}^5$. Then

$$\tilde{\sigma}_Q^2(\tilde{\mathcal{A}}) = \sum_{|n| \leq K|\ln \mathbf{r}|} \int_{\tilde{\Omega}_\mathcal{D}} \tilde{\mathcal{A}} \left(\tilde{\mathcal{A}} \circ \tilde{\mathcal{F}}_{Q,\mathbf{r}}^n \right)^T d\tilde{\mu}_\mathcal{D} + \mathcal{O}(\mathbf{r}^3).$$

Next we prove that for each $n \neq 0$

$$(\text{A.45}) \qquad \left| \int_{\tilde{\Omega}_\mathcal{D}} \tilde{\mathcal{A}} \left(\tilde{\mathcal{A}} \circ \tilde{\mathcal{F}}_{Q,\mathbf{r}}^n \right)^T d\tilde{\mu}_\mathcal{D} \right| \leq \text{Const}\, \mathbf{r}^2$$

By the time symmetry it is enough to consider $n > 0$. Let $\tilde{\tilde{\mathcal{A}}} = \tilde{\mathcal{A}} \circ \tilde{F}_Q^{-1}$. Then (A.45) is equivalent to

$$\text{(A.46)} \qquad \left| \int_{\tilde{\Omega}_{\mathcal{D}}} \tilde{\tilde{\mathcal{A}}} \left(\tilde{\mathcal{A}} \circ \tilde{\mathcal{F}}_{Q,\mathbf{r}}^{n-1} \right)^T d\tilde{\mu}_{\mathcal{D}} \right| \leq \text{Const}\, \mathbf{r}^2.$$

Consider domains

$$\tilde{\Pi}_{\mathbf{r}} := \{x \colon \tilde{\mathcal{A}} \neq 0\}, \quad \text{and} \quad \tilde{\tilde{\Pi}}_{\mathbf{r}} := \{x \colon \tilde{\tilde{\mathcal{A}}} \neq 0\}$$

in the space $\tilde{\Omega}_{\mathcal{D}}$. Observe that $\tilde{\Pi}$ consists of points that are about to collide with $\mathbb{B}_0$, and $\tilde{\tilde{\Pi}}$ consists of points that just collided with $\mathbb{B}_0$. Thus $\tilde{\Pi}$ is a finite union of narrow strips of width $\mathcal{O}(\mathbf{r})$ stretching along some s-curves, while $\tilde{\tilde{\Pi}}$ is a finite union of strips of width $\mathcal{O}(\mathbf{r})$ stretching along some u-curves, see Fig. 2. Observe also that $\cap_{\mathbf{r}>0} \tilde{\Pi}_{\mathbf{r}}$ is a finite union of s-curves $\tilde{\gamma}_i \subset \tilde{\Omega}_{\mathcal{D}}$ (consisting of points $x \in \tilde{\Omega}$ whose trajectories run straight into the point Q), and $\cap_{\mathbf{r}>0} \tilde{\tilde{\Pi}}_{\mathbf{r}}$ is a finite union of u-curves $\tilde{\tilde{\gamma}}_i \subset \tilde{\Omega}_{\mathcal{D}}$ whose trajectories come straight from the point Q.

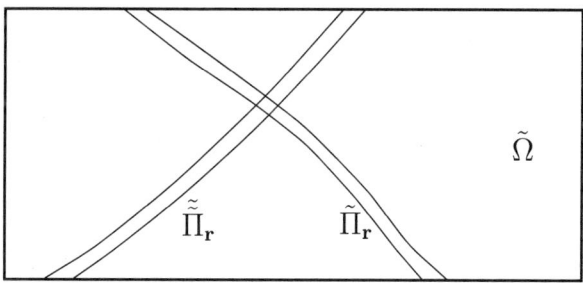

FIGURE 2. Two narrow strips in $\tilde{\Omega}$

Now the estimate (A.46) is obvious for $n = 1$. For $n > 1$ we apply the growth lemma 4.10. The domain $\tilde{\tilde{\Pi}}$ can be easily foliated by H-curves of length $\mathcal{O}(1)$ (independent of $\mathbf{r}$). If the foliation is smooth enough, the conditional measures on its fibers will have homogeneous densities, cf. Section 4.3, thus they become standard pairs. Then Lemma 4.10 implies that at any time $n > 1$ the images of those fibers will consist, on average, of curves of length $\mathcal{O}(1)$. Thus the fraction of that image intersecting $\tilde{\Pi}$ will be $< \text{Const}\,\mathbf{r}$. Integrating over all the fibers we obtain

$$\tilde{\mu}_{\mathcal{D}}\left(x \in \tilde{\tilde{\Pi}} \colon \tilde{\mathcal{F}}_{Q,\mathbf{r}}^{n-1}(x) \in \tilde{\Pi} \right) \leq \text{Const}\,\mathbf{r}\,\tilde{\mu}_{\mathcal{D}}(\tilde{\tilde{\Pi}}) \leq \text{Const}\,\mathbf{r}^2$$

This implies (A.45).

It remains to compute the $n = 0$ term

$$\text{(A.47)} \qquad \int_{\tilde{\Omega}_\mathcal{D}} \tilde{\mathcal{A}} \tilde{\mathcal{A}}^T \, d\tilde{\mu}_\mathcal{D} = L_\mathbf{r} \int_{\Omega^*_{Q,\mathbf{r}}} \mathcal{A} \mathcal{A}^T \, d\mu_{Q,\mathbf{r}}$$

where $\Omega^*_{Q,\mathbf{r}} = \partial \mathcal{P}(Q) \times [-\pi/2, \pi/2]$ is the collision space of the disk $\mathcal{P}(Q)$. The measure $\mu_{Q,\mathbf{r}}$ has density $c^{-1} \cos\varphi \, dr \, d\varphi$ in the coordinates r, φ introduced in Section 4.2, where $c = 2\,\text{length}(\partial \mathcal{D}) + 4\pi \mathbf{r}$ is the normalization factor. For convenience, we replace the arclength parameter r on $\partial \mathcal{P}(Q)$ with the angular coordinate $\psi \in [0, 2\pi)$, then we get the ψ, φ coordinates on $\Omega^*_{Q,\mathbf{r}}$ and $d\mu_{Q,\mathbf{r}} = c^{-1} \mathbf{r} \cos\varphi \, d\psi \, d\varphi$.

Due to an obvious rotational symmetry, the matrix (A.47) is a scalar multiple of the identity matrix, so it is enough to compute its first diagonal entry. The first component of the vector function $\mathcal{A}$ is $2 \cos\psi \cos\varphi$, hence the first diagonal entry of (A.47) is

$$\frac{\mathbf{r}}{2\,\text{length}(\partial\mathcal{D})} \int_{-\pi/2}^{\pi/2} \int_0^{2\pi} 4 \cos^2\psi \cos^3\varphi \, d\psi \, d\varphi = \frac{8\pi \mathbf{r}}{3\,\text{length}(\partial\mathcal{D})}$$

This completes the proof of (A.43), and hence that of (2.20). □

We have established the necessary result (2.20), but, as T. Spencer pointed out to us, the importance of the diffusion matrix in physics justified further analysis to obtain the more refined formula (A.42), which we do next.

The proof of (A.42) requires more accurate calculation of the integral (A.46). A crucial observation is that for fixed n the intersection $\tilde{\Pi}_\mathbf{r} \cap \tilde{\mathcal{F}}^{-n+1} \tilde{\Pi}_\mathbf{r}$ tends to concentrate around finitely many points, which we call *core points* and denote by

$$\bigcap_{\mathbf{r} > 0} \text{clos}\left(\tilde{\Pi}_\mathbf{r} \cap \tilde{\mathcal{F}}^{-n+1} \tilde{\Pi}_\mathbf{r}\right) = \{x_1^{(n)}, \ldots, x_{k_n}^{(n)}\}$$

(here clos(A) means the closure of A). These points corresponds to billiard trajectories under the map $\mathcal{F}_{Q,0}$ (on the table with a "degenerate" scatterer $\mathbb{B}_0$ of radius $\mathbf{r} = 0$) starting and ending at $\mathbb{B}_0 = \{Q\}$. We distinguish *non-singular core points* (which do not experience tangential collisions between their visits to Q) and other (*singular*) core points. The values of Z_Q in (A.42) can be computed by using the trajectories of the core points.

Our proof of (A.42) consists of five steps.

Step 1. We show that

$$\tilde{\mathcal{I}}_n := \int_{\tilde{\Omega}_\mathcal{D}} \tilde{\mathcal{A}} \left(\tilde{\mathcal{A}} \circ \tilde{\mathcal{F}}_{Q,\mathbf{r}}^n \right)^T d\tilde{\mu}_\mathcal{D} = a_n(Q)\,\mathbf{r}^2 + b_n(Q,\mathbf{r}) + \mathcal{O}(\mathbf{r}^{2+\delta})$$

for some $\delta > 0$. Here the first term, $a_n(Q)\,\mathbf{r}^2$, corresponds to the contribution of non-singular core points, the second term describes the contribution of singular core points, and the third term accounts for the trajectories hitting $\mathbb{B}_0$ more than once before time n and for non-linear effects.

Step 2. We establish an *a priori* bound $b_n(Q,\mathbf{r}) \leq \mathrm{Const}\,\theta^n\,\mathbf{r}^2$ where $\theta < 1$ for the contribution of the singular core points.

Step 3. From Steps 1 and 2 and the estimate (A.44) we conclude that

$$a_n(Q) = \frac{\tilde{\mathcal{I}}_n}{\mathbf{r}^2} + \mathcal{O}\left(\theta^n + \mathbf{r}^\delta\right) = \mathcal{O}\left(\frac{\theta^n}{\mathbf{r}^4} + \mathbf{r}^\delta\right).$$

Since the left hand side does not depend on $\mathbf{r}$, we can optimize our bound in $\mathbf{r}$ to get $a_n(Q) = \mathcal{O}(\tilde{\theta}^n)$ for some $\tilde{\theta} < 1$.

Step 4. We fix n and show that

$$\frac{b_n(Q,\mathbf{r})}{\mathbf{r}^2} \to b_n(Q) \quad \text{as } n \to \infty.$$

We should note that $b_n(Q)$ describes the contribution of singular core points, and the existence of such points is a "codimension one event" (there are only countably many orbits starting from Q and making a tangential collision), and there is no reason for them to pass through Q again). Thus for most Q we expect $b_n(Q) = 0$. However, since Q varies over a two-dimensional domain, we do expect a non-zero contribution for some exceptional values of Q.

Step 5. The estimates of steps 2–4 and the dominated convergence theorem imply

$$\sum_n \frac{\tilde{\mathcal{I}}_n}{\mathbf{r}^2} \to \sum_n \left[a_n(Q) + b_n(Q) \right].$$

Steps 3 and 5 are self-explanatory. We now describe estimates in Steps 1, 2 and 4 in more detail. First we compute $a_n(Q)$. For every point $q \in \partial \mathcal{D}$ with coordinate r we denote by $\mathbf{e}_Q(r)$ the unit vector pointing from q to Q, and by $d_Q(r)$ the distance from q to Q. Let $x_i^{(n)} = (r_i^{(n)}, \varphi_i^{(n)})$ be a nonsingular core point. In its vicinity, i.e. for $\left| r - r_i^{(n)} \right| < \varepsilon$, we have

$$\tilde{\Pi}_\mathbf{r} = \left\{ |\varphi - \varphi^*(r)| < \sin^{-1}\left(\frac{\mathbf{r}}{d(r)} \right) \right\},$$

where $\varphi^*(r)$ denotes the reflection angle of the unique trajectory arriving at r straight from the point Q. For small $\mathbf{r}$ we can approximate

$$\sin^{-1}\left(\frac{\mathbf{r}}{d(r)}\right) = \frac{\mathbf{r}}{d(r)} + \mathcal{O}(\mathbf{r}^3)$$

and so $\tilde{\tilde{\mathcal{A}}} = \tilde{\tilde{u}}\left(r, \frac{\varphi - \varphi^*(r)}{\mathbf{r}}\right) + \mathcal{O}(\mathbf{r}^2)$, where $\|\tilde{\tilde{u}}(r,s)\| = 2\sqrt{1 - [d(r)s]^2}$, and $\tilde{\tilde{u}}$ makes angle $\pi - \sin^{-1}(d(r)s)$ with the vector $\mathbf{e}_Q(r)$. Similar formulas apply to $\tilde{\mathcal{A}}$.

Observe that the set $\tilde{\mathcal{F}}_Q^{-n}\tilde{\Pi}_\mathbf{r} \cap \tilde{\tilde{\Pi}}_\mathbf{r}$ consists of three (not necessarily disjoint) parts:

(1) Vicinities of nonsingular core points that come back to Q for the first time in exactly n collisions;
(2) Vicinities of nonsingular core points that come back to Q more than once in the course of n collisions;
(3) Vicinities of singular core points, more precisely, orbits passing in the $\mathbf{r}$–neighborhood of tangential collisions before returning to Q.

We claim that, at least for large n, the main contribution to the integral $\tilde{\mathcal{I}}_n$ comes from orbits of the first type. To make this statement precise we denote by $I_2(n)$ and $I_3(n)$ the contribution of type 2 and type 3 orbits, respectively. To estimate $I_2(n)$ we observe that for fixed $k < n$ the set of points having collision with $\mathbb{B}_0$ immediately before the k-th return has measure at most $\text{Const}\,\mathbf{r}^2$. The images of these points can be foliated by u-curves of length $\mathcal{O}(1)$, so the contribution of such points is bounded by

$$|I_2(k,n)| \leq \text{Const}\,\mathbf{r}^3$$

Summation over $k = 1, \ldots, n$ gives

$$|I_2(n)| \leq \text{Const}\,\mathbf{r}^3 |\ln \mathbf{r}|.$$

Next we turn to the type 3 orbits. Let x be such an orbit and $k \in [1, n]$ denote the first moment of time when its image passes in the $\mathbf{r}$–neighborhood of a tangential collision. Assume first that $k \leq n/2$. Again, the measure of the set of all such orbits is $\text{Const}\,\mathbf{r}^2$, and it can be foliated by u-curves of length $\mathcal{O}(\mathbf{r})$. Denote by $\gamma_k(x)$ the u-curve containing the image of our point x. Consider now the images of $\gamma_k(x)$ at time $\frac{3n}{4}$ and denote by $\tilde{r}_{3n/4}(x)$ the distance from the corresponding image of our point x to the nearest endpoint of the u-curve it belongs to. Pick λ slightly larger than 1, then we have two cases:

(1) $\tilde{r}_{3n/4}(x) < \lambda^n \mathbf{r}$. By Lemma 4.10, if λ is sufficiently close to 1, then the measure of all such points is less than $\text{Const}\, \theta^n \mathbf{r}^2$ with some $\theta < 1$.

(2) $\tilde{r}_{3n/4}(x) \geq \lambda^n \mathbf{r}$. In this case we can again use the growth lemma 4.10 to conclude that the conditional probability of later hitting $\mathbb{B}_0$ (i.e. coming to $\tilde{\Pi}_\mathbf{r}$) is at most

$$\text{Const}\, \frac{\mathbf{r}}{\lambda^n \mathbf{r}} = \frac{\text{Const}}{\lambda^n},$$

and so the total contribution of such orbits is at most $\text{Const}\, \mathbf{r}^2/\lambda^n$. Summation over $k \leq n/2$ gives the combined contribution that can be expressed as $\text{Const}\, n \mathbf{r}^2 (\theta^n + 1/\lambda^n)$.

The case $k \geq n/2$ can be reduced to the previous one by using the time reversal property of billiard dynamics. Hence $|I_3(n)| \leq \text{Const}\, \tilde{\theta}^n$ for some $\tilde{\theta} < 1$, thus establishing the estimate claimed in Step 2.

Let us now compute the contribution of type 1 orbits. We only outline the argument leaving the (elementary but lengthy) estimates of some higher-order terms out. Let $x = x_i^{(n)} = \tilde{\mathcal{F}}_Q^{-n} \tilde{\gamma}_{j'} \cap \tilde{\gamma}_{j''}$ be a nonsingular core point. Choose a frame in $\mathcal{T}_x\Omega$ consisting of a unit vector tangent to $\tilde{\gamma}_{j''}$ and $\frac{\partial}{\partial \varphi}$. Consider a frame in $\mathcal{T}_{\tilde{\mathcal{F}}_Q^n x}\Omega$ consisting of a unit vector tangent to $\tilde{\gamma}_{j'}$ and $\frac{\partial}{\partial \varphi}$. Denote by $\zeta(x) = D_x \mathcal{F}_Q^n$ the 2×2 matrix of the derivative of the map $\mathcal{F}_Q^n$ in these frames. Foliate a neighborhood of x by curves

$$\sigma_c = \left\{ \frac{\varphi - \varphi^*(r)}{\mathbf{r}} = c \right\}.$$

Then $\tilde{\tilde{\mathcal{A}}}$ is approximately constant on each such curve. When $\mathbf{r}$ is sufficiently small the image $\tilde{\mathcal{F}}_Q^n \sigma_c$ intersects $\tilde{\Pi}_\mathbf{r}$ in a curve which is close to the straight line $D_x \tilde{\mathcal{F}}_Q^n(\mathcal{T}_x \tilde{\gamma})$ and the length of its preimage is about $2\mathbf{r}/\zeta_{21}(x)$. On the curve σ_c, we have $\tilde{\tilde{\mathcal{A}}} = \tilde{u}(r,c) + \mathcal{O}(\mathbf{r})$ so the average value of $\tilde{\tilde{\mathcal{A}}}\tilde{\mathcal{A}}$ over the above intersection is

$$\tilde{\tilde{u}}(r(x),c) \int \tilde{u}\bigl(r(F^n x), \tilde{c}\bigr)\, d\tilde{c} + \mathcal{O}(\mathbf{r}).$$

Observe that the integral here is a vector parallel to $\mathbf{e}_Q(r(\mathcal{F}_Q^n x))$. To compute the magnitude of this vector consider an angular coordinate ψ on $\mathbb{B}_0$ such that its value $\psi = 0$ corresponds to the direction toward the point $r(\mathcal{F}_Q^n x)$. In this coordinate, possible angles of collision range from $-\frac{\pi}{2} + \mathcal{O}(\mathbf{r})$ to $\frac{\pi}{2} + \mathcal{O}(\mathbf{r})$. The incoming vector is close to $(-1, 0)$,

so the outgoing vector is close to $\bigl(\cos(2\psi), \sin(2\psi)\bigr)$, hence
$$\tilde{u} \approx \bigl(\cos(2\psi) + 1, \sin(2\psi)\bigr) = \bigl(2(1 - \sin^2 \psi), \sin(2\psi)\bigr).$$
Finally the incoming vector makes angle close to $\mathbf{r} \sin \psi / d(r(\mathcal{F}^n x))$. Hence the length of the average momentum change is close to
$$\frac{\int_{-\pi/2}^{\pi/2} (1 - \sin^2 \psi) \, d \sin \psi}{\int_{-\pi/2}^{\pi/2} d \sin \psi} = \frac{4}{3}.$$

Next, averaging over c and using the density of the invariant measure $d\mu = [\text{length}(\mathcal{D})]^{-1} \cos \varphi \, dr \, d\varphi$ gives the total contribution of the nonsingular core point x, which we denote by $Z_Q(x) \, \mathbf{r}^2 + \mathcal{O}(\mathbf{r}^3)$, where
$$Z_Q(x) = - \frac{64 \cos \varphi^*(x)}{9 d(x) d(\mathcal{F}^n x) \zeta_{21}(x) \, \text{length}(\partial \mathcal{D})} \, \mathbf{e}_Q(r(x)) \otimes \mathbf{e}_Q\bigl(r(\mathcal{F}^n x)\bigr).$$
Thus
$$\sum_n a_n(Q) = \sum_x Z_Q(x),$$
where the sum is taken over all nonsingular core points of type 1. This completes estimate claimed in Step 1.

It remains to compute the contribution of type 3 orbits around singular core points (which only occur for some exceptional values of Q, as we explained above). This can be done similarly to $a_n(Q)$, except now the two parts of $\tilde{\mathcal{F}}_Q^{-n} \tilde{\Pi}_\mathbf{r} \cap \tilde{\Pi}_\mathbf{r}$ separated by the discontinuity curve have to be treated differently. The part experiencing an almost grazing collision makes no contribution to $b_n(Q)$, since $\zeta_{21} = \infty$. The contribution of the part avoiding the grazing collision is computed similarly to the type 1 orbits, but now the image of $\mathcal{F}_Q^n \sigma_c$ will be cut by the singularity curve, which needs to be approximated by its tangent line. The resulting expression for $b_n(Q)$ is not very useful, so we do not include it here. $\square$

APPENDIX B

Growth and distortion in dispersing billiards

B.1. Regularity of H-curves. It is known that for C^r smooth uniformly hyperbolic maps, such as Anosov diffeomorphisms, unstable manifolds are uniformly C^r smooth and the conditional densities of the SRB measure on unstable manifolds are uniformly C^{r-1} smooth.

In the case of billiards, the collision map $T\colon \Omega \to \Omega$ is C^r smooth whenever the table border $\partial \mathcal{D}$ is C^{r+1} smooth. Then we also have the C^r smoothness of unstable manifolds and the C^{r-1} smoothness of SRB densities, but not uniformly over the space Ω, since the corresponding derivatives explode near the singularities. Here we establish certain uniform bounds on the corresponding first and second derivatives.

Let $W_0 \subset \Omega$ be an H-curve, $x_0 = (r_0, \varphi_0) \in W_0$, and for every $n \geq 1$ denote by W_n the H-component of $\mathcal{F}^n(W_0)$ containing the point $x_n = (r_n, \varphi_n) = \mathcal{F}^n(x_0)$. In the r, φ coordinates, the curve W_n is a function $\varphi(r)$ and we denote its slope at the point x_n by $\Gamma_n = d\varphi/dr$. Recall that we use the metric (4.13) on u-curves, in which the norm of tangent vectors $dx = (dr, d\varphi)$ to u-curves satisfies

(B.1) $$\|dx\|^2 = (dr\cos\varphi)^2 + (d\varphi + \mathcal{K}\,dr)^2$$

Recall that by (4.14) $0 < c_1 \leq |dx/dr| \leq c_2 < \infty$ for some constants c_1, c_2. Let

$$\mathcal{J}_{W_i}\mathcal{F}^{-1}(x_i) = \left[\mathcal{J}_{W_{i-1}}\mathcal{F}(x_{i-1})\right]^{-1} = |dx_{i-1}|/|dx_i|$$

denote the Jacobian (the contraction factor) of the map $\mathcal{F}^{-1}\colon W_i \to W_{i-1}$ at the point x_i.

PROPOSITION B.1. *Suppose the boundary $\partial \mathcal{D}$ is of class C^3 and $|d\Gamma_0/dx_0| \leq C_0$ for some $C_0 > 0$ and all $x_0 \in W_0$. Then there is a constant $C > 0$ such that for all $n \geq 1$*

(B.2) $$\left|\frac{d\Gamma_n}{dx_n}\right| \leq C$$

and

(B.3) $$\left|\frac{d\ln \mathcal{J}_{W_n}\mathcal{F}^{-1}(x_n)}{dx_n}\right| \leq \frac{C}{|W_n|^{2/3}}$$

Suppose, in addition, that the boundary ∂D is of class C^4 and moreover $|d^2\Gamma_0/dx_0^2| \leq C_0$. Then for all $n \geq 1$

$$\left| d^2\Gamma_n/dx_n^2 \right| \leq C \tag{B.4}$$

and

$$\left| \frac{d^2 \ln \mathcal{J}_{W_n} \mathcal{F}^{-1}(x_n)}{dx_n^2} \right| \leq \frac{C}{|W_n|^{4/3}} \tag{B.5}$$

The first part of this proposition (related to a C^3 boundary) is known – full proofs are provided in [19], even for a more general class of billiards, where a small external field is permitted. The second part related to a C^4 boundary is new.

Before giving a proof, we derive a corollary. Let ρ_0 denote a density on W_0 and ρ_n the induced density on W_n:

$$\rho_n(x_n) = \rho_0(x_0)\, \mathcal{J}_{W_n}\mathcal{F}^{-n}(x_n) \tag{B.6}$$

COROLLARY B.2. *Suppose the boundary ∂D is of class C^3 and $|d\Gamma_0/dx_0| \leq C_0$ and $|d\ln \rho_0/dx_0| \leq C_0$. Then there is a constant $C > 0$ such that for all $n \geq 1$*

$$\left| \frac{d\ln \rho_n}{dx_n} \right| \leq \frac{C}{|W_n|^{2/3}} \tag{B.7}$$

Suppose, in addition, that the boundary ∂D is of class C^4 and moreover $|d^2\Gamma_0/dx_0^2| \leq C_0$. Then for all $n \geq 1$

$$\left| \frac{d^2 \ln \rho_n}{dx_n^2} \right| \leq \frac{C}{|W_n|^{4/3}} \tag{B.8}$$

Proof. This follows by logarithmic differentiation of (B.6) and using the following simple estimate:

$$\frac{1}{|W_i|^{2/3}} \left| \frac{dx_i}{dx_n} \right| = \frac{\mathcal{J}_{W_n}\mathcal{F}^{i-n}(x_n)}{|W_i|^{2/3}} \leq \frac{\text{Const }\vartheta^{\frac{n-i}{3}}}{|W_n|^{2/3}}$$

where $\vartheta^{-1} > 1$ denotes the minimal factor of expansion of u-curves. □

Proof of Proposition B.1. The curve $W = W_i$ corresponds to a family of trajectories of the billiard flow Φ^t. Let t_i be the reflection time for the trajectory of the point x_i. The tangent vector $(dr_i, d\varphi_i) \in \mathcal{T}_x W$ corresponds to a (time-dependent) tangent vector (dq_t, dv_t) to the orthogonal cross-section of that family, as it was shown in Chapter 4 (note that both dq_t and dv_t here are perpendicular to the velocity vector v_t of the family, since $M = \infty$). Denote by $\mathcal{B}_t = |dv_t|/|dq_t| > 0$ the curvature of the family.

The following facts are standard in billiard theory [**16, 18**] and can be obtained directly:

(B.9) $\quad\quad \frac{d}{dt} dq_t = dv_t, \quad \frac{d}{dt} dv_t = 0, \quad \frac{d}{dt}(\mathcal{B}_t^{-1} - t) = 0$

(provided t is not a moment of collision) and

(B.10) $\quad\quad |dq_{t_i^+}| = |dq_{t_i^-}|, \quad \mathcal{B}_{t_i^+} = \mathcal{B}_{t_i^-} + \frac{2\mathcal{K}(r_i)}{\cos \varphi_i}$

at a moment of collision (here $\mathcal{K}(r) > 0$ denotes the curvature of $\partial \mathcal{D}$ at the point r, and t_i^-, t_i^+ refer to the precollisional and postcollisional moments, respectively). One can see that

$$c_1 \leq \mathcal{B}_{t_i^-} \leq c_2, \quad c_1 \leq \cos \varphi_i \, \mathcal{B}_{t_i^+} < c_2$$

for some constants $0 < c_1 < c_2 < \infty$. Note that $\tilde{\mathcal{B}}_i := \mathcal{B}_{t_i^-}$ remains uniformly bounded. In fact, all our troubles come from the unbounded factor $1/\cos \varphi_i$ in (B.10).

LEMMA B.3. *Suppose the boundary $\partial \mathcal{D}$ is of class C^3 and also $|d\tilde{\mathcal{B}}_0/dx_0| \leq C_0'/\cos^2 \varphi_0$ for some $C_0' > 0$ and all $x_0 \in W_0$. Then there is a constant $C' > 0$ such that for all $n \geq 1$*

(B.11) $\quad\quad \left| d\tilde{\mathcal{B}}_n/dr_n \right| \leq C'$

Suppose, in addition, that the boundary $\partial \mathcal{D}$ is of class C^4 and moreover $|d^2 \tilde{\mathcal{B}}_0/dx_0^2| \leq C_0''/\cos^2 \varphi_0$. Then for all $n \geq 1$

(B.12) $\quad\quad \left| d^2 \tilde{\mathcal{B}}_n/dr_n^2 \right| \leq C''$

We postpone the proof of the lemma and complete the proof of Proposition B.1 first.

The slope $\Gamma_n = d\varphi_n/dr_n$ of the curve W_n satisfies

(B.13) $\quad\quad \Gamma_n = \tilde{\mathcal{B}}_n \cos \varphi_n + \mathcal{K}(r_n)$

hence

$$\frac{d\Gamma_n}{dx_n} = \frac{\frac{d\tilde{\mathcal{B}}_n}{dr_n} \cos \varphi_n - \tilde{\mathcal{B}}_n \Gamma_n \sin \varphi_n + \frac{d\mathcal{K}(r_n)}{dr_n}}{[\cos^2 \varphi_n + (\Gamma_n + \mathcal{K}(r_n))^2]^{1/2}}$$

(the denominator equals $|dx_n/dr_n|$ according to (B.1)). It is easy to see that our assumption $|d\Gamma_0/dx_0| \leq C_0$ implies $|d\tilde{\mathcal{B}}_0/dx_0| \leq C_0'/\cos \varphi_0$ for some constant $C_0' > 0$. Now (B.2) follows from (B.11).

Differentiating further gives an expression for $d^2 \Gamma_n/dx_n^2$ (we leave it to the reader), and it shows that our assumption $|d^2 \Gamma_0/dx_0^2| \leq C_0$ implies $|d^2 \tilde{\mathcal{B}}_0/dx_0^2| \leq C_0''/\cos^2 \varphi_0$ for some $C_0'' > 0$. Now (B.4) follows from (B.12).

It remains to prove (B.3) and (B.5). To compute the Jacobian $\mathcal{J}_{W_n}\mathcal{F}^{-1}(x_n) = |dx_{n-1}|/|dx_n|$ we note that by (B.10)
$$|dx_n|^2 = |dq_{t_n^+}|^2 + |dv_{t_n^+}|^2 = |dq_{t_n^-}|^2(1+\mathcal{B}_{t_n^+}^2)$$
hence
$$\mathcal{J}_{W_n}\mathcal{F}^{-1}(x_n) = \frac{|dq_{t_{n-1}^+}|}{|dq_{t_n^-}|}\left[\frac{1+\mathcal{B}_{t_{n-1}^+}^2}{1+\mathcal{B}_{t_n^+}^2}\right]^{1/2}$$
where
$$(B.14) \qquad \frac{|dq_{t_{n-1}^+}|}{|dq_{t_n^-}|} = \frac{1}{1+(t_n-t_{n-1})\mathcal{B}_{t_{n-1}^+}}$$

It follows that
$$2\ln\mathcal{J}_{W_n}\mathcal{F}^{-1}(x_n) = -\ln\left[1+\left(\tilde{\mathcal{B}}_n + \frac{2\mathcal{K}(r_n)}{\cos\varphi_n}\right)^2\right]$$
$$+ \ln\left[\tilde{\mathcal{B}}_n^2 + \left(1 - (t_n-t_{n-1})\tilde{\mathcal{B}}_n\right)^2\right]$$

We also note that
$$(B.15) \qquad dt_n/dr_n = \pm\sin\varphi_n$$
and $d^2t_n/dr_n^2 = \pm\Gamma_n\cos\varphi_n$ (where the sign depends on the orientation of the tangent vector $(dr_n, d\varphi_n)$). Now one can differentiate $\ln\mathcal{J}_{W_n}\mathcal{F}^{-1}(x_n)$ directly and use Lemma B.3 to derive bounds
$$\left|\frac{d\ln\mathcal{J}_{W_n}\mathcal{F}^{-1}(x_n)}{dx_n}\right| \leq \frac{\text{Const}}{\cos\varphi_n} \quad \text{and} \quad \left|\frac{d^2\ln\mathcal{J}_{W_n}\mathcal{F}^{-1}(x_n)}{dx_n^2}\right| \leq \frac{\text{Const}}{\cos^2\varphi_n}$$
which imply (B.3) and (B.5) due to (4.17). □

Proof of Lemma B.3. Our argument has an inductive character. Observe that
$$(B.16) \qquad \tilde{\mathcal{B}}_n = \frac{1}{t_n - t_{n-1} + [\tilde{\mathcal{B}}_{n-1} + 2\mathcal{K}(r_{n-1})/\cos\varphi_{n-1}]^{-1}}$$
$$= \frac{1}{t_n-t_{n-1}} - \frac{1}{(t_n-t_{n-1})^2\left(\tilde{\mathcal{B}}_{n-1} + \frac{2\mathcal{K}(r_{n-1})}{\cos\varphi_{n-1}} + \frac{1}{t_n-t_{n-1}}\right)}.$$

Next,
$$\frac{|dr_{n-1}|}{|dr_n|} = \frac{|dq_{t_{n-1}}|/\cos\varphi_{n-1}}{|dq_{t_n}|/\cos\varphi_n} = \frac{\cos\varphi_n}{w_{n-1}}$$
where
$$w_{n-1} = 2\mathcal{K}(r_{n-1})(t_n - t_{n-1}) + \cos\varphi_{n-1}\bigl(1+(t_n-t_{n-1})\tilde{\mathcal{B}}_{n-1}\bigr)$$

Note that w_{n-1} is uniformly bounded above and below:
$$0 < w_{\min} \leq w_{n-1} \leq w_{\max} < \infty$$

Now a direct differentiation of (B.16) using (B.13) and (B.15) gives
$$\left|\frac{d\tilde{\mathcal{B}}_n}{dr_n}\right| = \theta_{n-1}^2 \theta_n \frac{w_n}{w_{n-1}} \left|\frac{d\tilde{\mathcal{B}}_{n-1}}{dr_{n-1}}\right| + \mathcal{R}'$$

where
$$\theta_{n-1} = \frac{|dq_{t_{n-1}^+}|}{|dq_{t_n^-}|} = \frac{1}{1 + (t_n - t_{n-1})\,\mathcal{B}_{t_{n-1}^+}} \leq \theta_{\max}$$

with
$$\theta_{\max} := \frac{1}{1 + L_{\min}\mathcal{K}_{\min}} < 1$$

(here $L_{\min}$ is the minimum free path between collisions), and the remainder term $\mathcal{R}'$ is uniformly bounded, $|\mathcal{R}'| \leq \mathcal{R}'_{\max}$. It is now easy to see that
$$\left|\frac{d\tilde{\mathcal{B}}_n}{dr_n}\right| \leq \frac{w_{\max}}{w_{\min}}\left(\frac{\mathcal{R}'_{\max}}{1 - \theta_{\max}^3} + \theta_0^2 \left|\frac{d\tilde{\mathcal{B}}_0}{dr_0}\right|\right)$$

Also note that $\theta_0^2 \leq \text{Const}\,\cos\varphi_0^2$. This proves (B.11).

Differentiating one more time and using (B.11) gives
$$\left|\frac{d^2\tilde{\mathcal{B}}_n}{dr_n^2}\right| = \theta_{n-1}^2 \theta_n^2 \frac{w_n^2}{w_{n-1}^2}\left|\frac{d^2\tilde{\mathcal{B}}_{n-1}}{dr_{n-1}^2}\right| + \mathcal{R}''$$

where $|\mathcal{R}''| \leq \mathcal{R}''_{\max}$. Now it is easy to see that
$$\left|\frac{d^2\tilde{\mathcal{B}}_n}{dr_n^2}\right| \leq \frac{w_{\max}^2}{w_{\min}^2}\left(\frac{\mathcal{R}''_{\max}}{1 - \theta_{\max}^4} + \theta_0^2\left|\frac{d^2\tilde{\mathcal{B}}_0}{dr_0^2}\right|\right)$$

This proves (B.12). $\square$

B.2. Invariant Section Theorem. Here we outline a proof of the general fact mentioned in Section 5.9. Let E^u be a family of unstable directions on an s-curve S, and $\Gamma(x) = d\varphi/dr > 0$ denote the slope of the E^u direction through the point $x \in S$. We say that E^u is Hölder continuous on S with exponent $a > 0$ and norm $L > 0$ if for all $x, y \in S$
$$|\Gamma(x) - \Gamma(y)| \leq L\,[\text{dist}(x,y)]^a$$

PROPOSITION B.4. *Let S be an s-curve such that $S_n = \mathcal{F}^n(S)$ is an s-curve for every $n = 1, \ldots, N$. If a family E^u on S is smooth enough, then the family $E_n^u = \mathcal{F}^n(E^u)$ on S_n is Hölder continuous with exponent $a = 1/2$ and norm $\leq C$ for all $n = 1, \ldots, N$, where $C > 0$ is a constant independent of N and S.*

Proof. We use the notation of the previous section. Let $x = (r, \varphi) \in S$ and $x_n = (r_n, \varphi_n) = \mathcal{F}^n(x) \in S_n$. Denote by $\Gamma_n(x_n)$ the slope of E_n^u direction at x_n. Due to (B.13)

$$\Gamma_n(x_n) = \tilde{\mathcal{B}}_n(x_n) \cos \varphi_n + \mathcal{K}(r_n)$$

where $\tilde{\mathcal{B}}_n(x_n)$ is the curvature of the incoming family of trajectories corresponding to the E_n^u direction at x_n. Since $\mathcal{K}(r_n)$ is uniformly C^1 smooth, it is enough to prove the Hölder continuity for $\tilde{\mathcal{B}}_n$ with a uniformly bounded norm. Let $x + dx = (r + dr, \varphi + d\varphi) \in S$ be a nearby point and $x_n + dx_n = (r_n + dr_n, \varphi_n + d\varphi_n) = \mathcal{F}^n(x + dx) \in S_n$. We will prove by induction on n that

(B.17) $$|\tilde{\mathcal{B}}_n(x_n + dx_n) - \tilde{\mathcal{B}}_n(x_n)| \leq \bar{C} u(x_n) |dr_n|^{1/2}$$

where $\bar{C} > 0$ is a large constant and $u(x)$ is a function (defined below), which is uniformly bounded:

$$0 < u_{\min} \leq u(x) \leq u_{\max} < \infty$$

(here $u_{\min}$ and $u_{\max}$ do not depend on N or S). Since the distance $|dr_n|$ is equivalent to our metric (4.15) on stable curves, the bound (B.17) implies Proposition B.4.

Now we prove (B.17). Due to (B.16)

$$\tilde{\mathcal{B}}_{n+1}(x_{n+1}) = \cfrac{1}{s(x_n) + \cfrac{1}{\tilde{\mathcal{R}}(x_n) + \tilde{\mathcal{B}}_n(x_n)}}$$

where $s(x)$ denotes the free path between the collision points at x and $\mathcal{F}(x)$, and $\tilde{\mathcal{R}}(x_n) = 2\mathcal{K}(r_n)/\cos \varphi_n$. For brevity, we will use notation $d\tilde{\mathcal{B}}_n = \tilde{\mathcal{B}}_n(x_n + dx_n) - \tilde{\mathcal{B}}_n(x_n)$, $ds_n = s(x_n + dx_n) - s(x_n)$, etc. Now elementary calculations give

$$|d\tilde{\mathcal{B}}_{n+1}| \leq \frac{|ds_n|}{[s(x_n)]^2} + \frac{|d\tilde{\mathcal{R}}_n| + |d\tilde{\mathcal{B}}_n|}{[\mathcal{J}_n(x_n)]^2}$$

where

$$\mathcal{J}_n(x_n) = 1 + s(x_n)[\tilde{\mathcal{R}}(x_n) + \tilde{\mathcal{B}}_n(x_n)]$$

Observe that $|ds_n| \leq |dr_n| + |dr_{n+1}|$ and

$$|d\tilde{\mathcal{R}}_n| \leq \frac{\text{Const } |dr_n|}{\cos^2 \varphi_n}$$

hence

$$\frac{|d\tilde{\mathcal{R}}_n|}{[\mathcal{J}_n(x_n)]^2} \leq \text{Const } |dr_n|$$

It is also easy to see that $|dr_n| \leq \text{Const} |dr_{n+1}|^{1/2}$, thus we obtain

(B.18) $$|d\tilde{\mathcal{B}}_{n+1}| \leq \text{Const} |dr_{n+1}|^{1/2} + |d\tilde{\mathcal{B}}_n|/[\mathcal{J}_n(x_n)]^2$$

Now pick a vector $0 \neq dr_n^u \in E_n^u(x_n)$ and put $dr_{n+1}^u = d\mathcal{F}(dr_n^u) \in E_{n+1}^u(x_{n+1})$. Using the notation of the previous section, we introduce $|dq_n| = |dr_n^u| \cos\varphi_n$ and $|dq_{n+1}| = |dr_{n+1}^u| \cos\varphi_{n+1}$, then

$$\mathcal{J}_n(x_n) = |dq_{n+1}|/|dq_n|$$

due to (B.14). The element of the Lebesgue measure $dm = dr\,d\varphi$ at the point x_n can be expressed by

$$dm(x_n) = |dr_n|\,|dr_n^u|\,w(x_n)$$

where

$$w(x_n) = d\varphi_n^u/dr_n^u - d\varphi_n/dr_n.$$

We note that $d\varphi_n^u/dr_n^u > 0$ and $d\varphi_n/dr_n < 0$. Then by (4.11) and (4.14), $w(x_n)$ is a function uniformly bounded above and below by positive constants:

$$0 < w_{\min} \leq w(x) \leq w_{\max} < \infty$$

cf. (4.11). Since the measure $d\mu = \cos\varphi\,dm$ is $\mathcal{F}$-invariant, we can write

$$|dr_n|\,|dr_n^u|\,w(x_n)\cos\varphi_n = |dr_{n+1}|\,|dr_{n+1}^u|\,w(x_{n+1})\cos\varphi_{n+1}$$

hence

$$|dr_n| = |dr_{n+1}|\,\mathcal{J}_n(x_n)\,\frac{w(x_{n+1})}{w(x_n)}$$

Now we set $u(x_n) = [w(x_n)]^{1/2}$ and use (B.18) and the inductive assumption (B.17) to get

$$|d\tilde{\mathcal{B}}_{n+1}| \leq \text{Const} |dr_{n+1}|^{1/2} + \frac{\bar{C}\,u(x_{n+1})\,|dr_{n+1}|^{1/2}}{[\mathcal{J}_n(x_n)]^{3/2}}$$

Since $\mathcal{J}_n(x_n) \geq 1 + L_{\min}\mathcal{K}_{\min} > 1$, we have

$$|d\tilde{\mathcal{B}}_{n+1}| \leq \bar{C}\,u(x_{n+1})\,|dr_{n+1}|^{1/2}$$

provided $\bar{C}$ is large enough. This proves (B.17) by induction. $\square$

B.3. The function space $\mathfrak{R}$.

Here we prove Lemma 3.1. Clearly, it is enough to prove it for $B_1 \equiv 1$. We use induction on n_A. For $n_A = 1$ the lemma reduces to the definition of $\mathfrak{R}$. For $n_A \geq 2$ we put $B = B_2 \circ \mathcal{F}^{n_A-2}$ and $A = B_2 \circ \mathcal{F}^{n_A-1}$. The Hölder continuity of A on the connected components of $\Omega \setminus \mathcal{S}_{n_A}$ follows from (3.8):

$$|A(x) - A(x')| = |B(\mathcal{F}(x)) - B(\mathcal{F}(x'))|$$
$$\leq K_B \left[\text{dist}(\mathcal{F}(x), \mathcal{F}(x'))\right]^{\alpha_B}$$
$$\leq K_B K_\mathcal{F}^{\alpha_B} \left[\text{dist}(x, x')\right]^{\alpha_B \alpha_\mathcal{F}}$$

It remains to estimate the local Lipschitz constant $\text{Lip}_x(A)$ defined by (3.9). First we note that $\text{Lip}_x(A) \leq \|D_x \mathcal{F}\| \text{Lip}_y(B)$, where $y = \mathcal{F}(x)$. The derivative $D_x \mathcal{F}$ is unbounded in the vicinity of $\mathcal{S}$, more precisely, on one side of $\mathcal{S}$ which corresponds to nearly grazing collisions, i.e. where y is close to $\partial\Omega$. Denote $d_1 = \text{dist}(x, \mathcal{S})$, $d_2 = \text{dist}(x, \mathcal{S}_{n_A} \setminus \mathcal{S})$, $d_3 = \text{dist}(y, \partial\Omega) \sim \pi/2 - |\varphi|$, where $y = (r, \varphi)$ in the notation of Section 4.2, and $d_4 = \text{dist}(y, \mathcal{S}_{n_A-1})$. All these distances are measured along some unstable curves, see Section 4.5, and $D_x \mathcal{F}$ attains its maximal expansion $\sim 1/\sqrt{d_1}$ along unstable curves through x, hence $d_3 \sim \sqrt{d_1}$. Note that $d := \text{dist}(x, \mathcal{S}_{n_A}) = \min\{d_1, d_2\}$. We now have two cases:

(a) If $d_1 < d_2$, then $d_3 < \text{Const } d_4$, hence we have $\text{dist}(y, \mathcal{S}_{n_A-1}) > \text{Const}^{-1} d_3$ and

$$\text{Lip}_x(A) \leq \frac{\text{Const}}{\sqrt{d_1}} \frac{\text{Const}}{d_3^{\beta_B}} \leq \frac{\text{Const}}{d^{(1+\beta_B)/2}}$$

(b) If $d_2 \leq d_1$, then $d_4 \leq \text{Const } d_3$ and $d_4 \sim d_2/\sqrt{d_1}$, hence

$$\text{Lip}_x(A) \leq \frac{\text{Const}}{\sqrt{d_1}} \frac{\text{Const}}{d_4^{\beta_B}} \leq \frac{\text{Const}}{d_1^{1/2-\beta_B/2} d_2^{\beta_B/2}} \leq \frac{\text{Const}}{d^{(1+\beta_B)/2}}$$

In either case we obtain the required estimate with $\beta_A = (1+\beta_B)/2$. Since $\beta_B < 1$, we have $\beta_A < 1$. Lemma 3.1 is proved. $\square$

Lastly, we prove Proposition 6.14. Its first claim follows from the fact that the configuration space of our system is a four dimensional domain bounded by cylindrical surfaces. To prove the Lipschitz continuity of $\mu_{Q,V}(d)$, consider two nearby points (Q, V) and (Q', V') and denote $h = \|Q - Q'\| + \|V - V'\|$. First assume that the light particle starts at a point (q, v) such that $q \in \partial \mathcal{D}$ and compare $d(Q, V, q, v)$ with $d(Q', V', q, v)$. It is convenient to use the coordinate frame moving with velocity vector V (in this frame, the disk Q, V is at rest). Then the light particle moves with velocity $v - V$ and the disk Q', V' moves with velocity $V' - V$. At the time of the next collision, the moving disk Q', V'

will be at distance $\mathcal{O}(h)$ from the fixed disk Q, V. One can check by direct inspection that the average difference $d(Q, V, q, v) - d(Q', V', q, v)$ is $\mathcal{O}(h)$. The other case $q \in \partial \mathcal{P}(Q)$ is easier, we leave it to the reader.
□

APPENDIX C

Distortion bounds for two particle system

Here we prove rather technical Propositions 4.6, 4.7 and Lemma 4.20 whose proofs were left out in Chapter 4.

We first outline our strategy. We have shown in Chapter 4 that unstable vectors $dx_t = (dq_t, dv_t, dQ_t, dV_t)$ grow with t through two alternating stages: free motion between collisions expands dq_t, while at collisions dv_t "jumps up". The resulting transformation of the tangent vectors is usually described by an operator-valued continued fraction [15, 16], and then distortion bounds can be proved by differentiating that fraction along unstable directions. This approach is convenient for completely hyperbolic billiards, because it treats all the components of unstable vectors equally. In our case, the components dq_t and dv_t expand uniformly, while dQ_t and dV_t change little and may not grow at all (effectively, we deal with a partially hyperbolic dynamics). We use a more explicit approach to prove distortion bounds here: pick two almost equal unstable vectors at nearby points on one u-curve and show that the images of these vectors have almost the same length at every iteration.

Let $x_0 = (Q_0, V_0, q_0, v_0) \in \Omega$ and $x_0' = (Q_0', V_0', q_0', v_0') \in \Omega$ be two nearby points that belong in one homogeneity section, say $\mathbb{H}_{m_0}$. Assume that for each $1 \leq i \leq n$ the points $x_i = (Q_i, V_i, q_i, v_i) = \mathcal{F}^i(x_0)$ and $x_i' = (Q_i', V_i', q_i', v_i') = \mathcal{F}^i(x_0')$ also belong in one homogeneity section, call it $\mathbb{H}_{m_i}$. We assume that for $0 \leq i \leq n$

$$\|Q_i - Q_i'\| \leq C \|q_i - q_i'\|/M$$

and

$$\|V_i - V_i'\| \leq C \|v_i - v_i'\|/M$$

where $C > 1$ is a large constant. Denote by (r_i, φ_i) and (r_i', φ_i') the coordinates of the points $\pi_0(x_i)$ and $\pi_0(x_i')$, respectively, and put

$$\delta_i = \sqrt{(r_i - r_i')^2 + (\varphi_i - \varphi_i')^2}$$

Assume that for all $i \leq n$

(C.1) $$\|q_i - q_i'\| \leq C\delta_i$$

and

(C.2) $$\|v_i - v'_i\| \leq C\delta_i$$

where $C > 1$ is a large constant. These assumptions hold, for example, when x_i and x'_i belong in one unstable curve (this follows from Propositions 4.1 and 4.4).

Let $dx_0 = (dQ_0, dV_0, dq_0, dv_0)$ be a postcollisional unstable vector at x_0, and $dx'_0 = (dQ'_0, dV'_0, dq'_0, dv'_0)$ a similar vector at x'_0. For $i \geq 1$, denote by $dx_i = (dQ_i, dV_i, dq_i, dv_i)$ and $dx'_i = (dQ'_i, dV'_i, dq'_i, dv'_i)$ their postcollisional images at the points x_i and x'_i, respectively. We say that the unstable vectors dx_i and dx'_i are $(\varepsilon_i, \tilde{\varepsilon}_i)$-close, if the following four bounds hold:

$$\|dq_i - dq'_i\| \leq \varepsilon_i \|dq_i\|,$$
$$\|dv_i - dv'_i\| \leq \varepsilon_i \|dv_i\|,$$
$$\|dQ_i - dQ'_i\| \leq \tilde{\varepsilon}_i \|dq_i\|/M,$$
$$\|dV_i - dV'_i\| \leq \tilde{\varepsilon}_i \|dv_i\|/M.$$

LEMMA C.1 (One-step distortion control). *Assume that the unstable vectors dx_0 and dx'_0 are $(\varepsilon_0, \tilde{\varepsilon}_0)$-close for some small $\varepsilon_0, \tilde{\varepsilon}_0 > 0$. Then their images dx_1 and dx'_1 will be $(\varepsilon_1, \tilde{\varepsilon}_1)$-close, where*

(C.3) $$\varepsilon_1 = \left(\varepsilon_0 + \frac{2\tilde{\varepsilon}_0}{M}\right)\left(1 + \frac{C}{\sqrt{M}}\right) + C\frac{\delta_0 + \delta_1}{\cos \varphi_1}$$

(C.4) $$\tilde{\varepsilon}_1 = \left(2\varepsilon_0 + \tilde{\varepsilon}_0\right)\left(1 + \frac{C}{\sqrt{M}}\right) + C\frac{\delta_0 + \delta_1}{\cos \varphi_1}$$

Here $C > 0$ is a large constant.

Proof. We first compare the precollisional vectors

$$dx_1^- = (dQ_1^-, dV_1^-, dq_1^-, dv_1^-)$$

and

$$(dx'_1)^- = ((dQ'_1)^-, (dV'_1)^-, (dq'_1)^-, (dv'_1)^-)$$

at the points x_1 and x'_1, respectively. Equation (4.1) and the triangle inequality imply

$$\|dq_1^- - (dq'_1)^-\| \leq \varepsilon_0 \|dq_0\| + s\,\varepsilon_0 \|dv_0\|$$
$$+ |s - s'|\,\|dv'_0\|$$

where s (resp., s') is the time between collisions at the points x_0 and x_1 (resp., x'_0 and x'_1). Due to Proposition 4.1 (d)–(e), the vectors dq_0

and dv_0 are almost parallel, for large M, hence we can combine the first two terms in the above bound:
$$\varepsilon_0 \|dq_0\| + s\,\varepsilon_0 \|dv_0\| \leq \varepsilon_0(1 + C/\sqrt{M})\,\|dq_1^-\|$$
Here and below we denote by $C = C(\partial \mathcal{D}, \mathbf{r}) > 0$ various constants. Next, it is a simple geometric fact that
$$\|sv - s'v'\| \leq \delta_0 + \delta_1 + \|Q_0 - Q_0'\| + \|Q_1 - Q_1'\|$$
and by the assumptions (C.1)–(C.2) we get
$$|s - s'| \leq C(\delta_0 + \delta_1)$$
Using Proposition 4.1 (g) gives
$$\text{(C.5)} \quad \|dq_1^- - (dq_1')^-\| \leq \varepsilon_0(1 + C/\sqrt{M})\,\|dq_1^-\| + C(\delta_0 + \delta_1)\,\|dq_1^-\|$$
Similarly,
$$\|dQ_1^- - (dQ_1')^-\| \leq \tilde{\varepsilon}_0(1 + C/\sqrt{M})\,\|dq_1^-\|/M$$
$$\text{(C.6)} \qquad\qquad + C(\delta_0 + \delta_1)\,\|dq_1^-\|/M$$
where the estimation of the last term involves Proposition 4.1 (c).

Now the postcollisional vectors dq_1 and dQ_1 depend on the precollisional vectors dq_1^- and dQ_1^- through certain reflection operators defined in terms of the normal vector n, see (4.3), (4.4). Since $\partial \mathcal{D}$ and $\partial \mathcal{P}(Q)$ are C^3 smooth, those reflection operators depend smoothly on x_1, with uniformly bounded derivatives, hence
$$\|dq_1 - dq_1'\| \leq \|dq_1^- - (dq_1')^-\| + 2\|dQ_1^- - (dQ_1')^-\|$$
$$+ C\delta_1\,\|dq_1^-\| + C\delta_1\,\|dQ_1^-\|$$
Applying (C.5)–(C.6) and Proposition 4.1 (b) gives
$$\|dq_1 - dq_1'\| \leq (\varepsilon_0 + 2\tilde{\varepsilon}_0/M)(1 + C/\sqrt{M})\,\|dq_1\|$$
$$+ C(\delta_0 + \delta_1)\,\|dq_1\|$$
Similarly,
$$\|dQ_1 - dQ_1'\| \leq \|dQ_1^- - (dQ_1')^-\| + 2\|dq_1^- - (dq_1')^-\|/M$$
$$+ C\delta_1\,\|dQ_1^-\| + C\delta_1\,\|dq_1^-\|/M$$
Applying (C.5)–(C.6) and Proposition 4.1 (b) gives
$$\|dQ_1 - dQ_1'\| \leq (2\varepsilon_0 + \tilde{\varepsilon}_0)\,(1 + C/\sqrt{M})\,\|dq_1\|/M$$
$$+ C(\delta_0 + \delta_1)\,\|dq_1\|/M$$

We now consider the velocity components dv and dV. They do not change between collisions. At collisions, these vectors are formed by certain reflection operators defined in terms of the normal n and acquire an addition involving the operator Θ^-, see Section 4.1. In those equations all the operators and vectors smoothly change with the point x_1 with bounded derivatives (see also (C.2)), except for the unbounded factor $\|w^+\|/\langle w^+, n\rangle$, which we later denoted by $1/\cos\varphi$. In what follows, we apply an elementary estimate for φ, φ' in the same homogeneous section:

(C.7) $$\left|\frac{1}{\cos\varphi_1} - \frac{1}{\cos\varphi_1'}\right| \leq \frac{|\varphi_1 - \varphi_1'|}{\cos\varphi_1 \cos\varphi_1'} \leq \frac{C\delta_1}{\cos^2\varphi_1}$$

Consider first the (simpler) case of a collision of the light particle with $\partial\mathcal{D}$. It follows from (4.2) that

$$\|\Theta^-\| = \frac{2\mathcal{K}\|v^+\|^2}{\langle v^+, n\rangle}$$

where all the vectors are taken at the point x_1. Thus, we obtain

(C.8) $$\|dv_1 - dv_1'\| \leq \|dv_0 - dv_0'\| + \|\Theta^-\|\|dq_1^- - (dq_1')^-\|$$
$$+ C\,\delta_1\,\|dv_1\| + C\|\Theta^-\|\|dq_1^-\|\delta_1/\cos\varphi_1.$$

By using (C.5), the sum of the first two terms on the right hand side in the above inequality can be bounded as follows:

$$A := \|dv_0 - dv_0'\| + \|\Theta^-\|\|dq_1^- - (dq_1')^-\|$$
$$\leq \varepsilon_0 \|dv_0\| + \varepsilon_0(1 + C/\sqrt{M})\,\|\Theta^-\|\,\|dq_1^-\| + C(\delta_0 + \delta_1)\|\Theta^-\|\|dq_1^-\|$$

It is also clear that the operator Θ^- attains its norm on the vectors perpendicular to $v^- = v_1^-$. By Proposition 4.1 the vector dq_1^- is almost perpendicular to v_1^-, thus $\|\Theta^-\|\|dq_1^-\| = (1 + \varkappa)\|\Theta^-(dq_1^-)\|$ with some $\varkappa = \mathcal{O}(1/\sqrt{M})$. Now we can combine the first two terms on the right hand side of the previous inequality as follows:

$$A' := \varepsilon_0\|dv_0\| + \varepsilon_0(1 + C/\sqrt{M})\,\|\Theta^-\|\,\|dq_1^-\|$$
$$\leq \varepsilon_0\|\mathbf{R}_n(dv_0)\| + \varepsilon_0(1 + C/\sqrt{M})\,\|\Theta^+(dq_1^+)\|$$
$$\leq \varepsilon_0(1 + C/\sqrt{M})\,\|dv_1\|$$

Finally, combining all our estimates gives

(C.9) $$\|dv_1 - dv_1'\| \leq \varepsilon_0(1 + C/\sqrt{M})\,\|dv_1\| + C(\delta_0 + \delta_1)\|dv_1\|/\cos\varphi_1$$

In the (more difficult) case of an interparticle collision we have a few extra terms in the main bound (C.8):

$$\|dv_1 - dv_1'\| \leq \cdots + 2\|dV_0 - dV_0'\| + C\delta_1\|dV_0\|$$
$$+ C\delta_1\|\Theta^-\|\|dQ_1^-\| + \|\Theta^-\|\|dQ_1^- - (dQ_1')^-\|$$

where $\cdots$ denote the terms already shown in (C.8). Now, the first new term above is bounded by

$$\text{(C.10)} \qquad 2\|dV_0 - dV_0'\| \leq 2\tilde{\varepsilon}_0 \|dv_0\|/M$$

The following two terms can be easily bounded and incorporated into the previous estimate (C.9). The last term $\|\Theta^-\| \|dQ_1^- - (dQ_1')^-\|$ can be bounded, with the help of (C.6), by

$$\tilde{\varepsilon}_0(1 + C/\sqrt{M}) \|\Theta^-\| \|dq_1^-\|/M + C(\delta_0 + \delta_1) \|\Theta^-\| \|dq_1^-\|/M$$

The second term in this expression can be easily incorporated into the previous estimate (C.9). To the first term we apply the same analysis of the operator Θ^- as was made in the case of a collision of the light particle with $\partial \mathcal{D}$, and then combine it with (C.10) and obtain the bound

$$2\tilde{\varepsilon}_0 \|dv_0\|/M + \tilde{\varepsilon}_0(1 + C/\sqrt{M}) \|\Theta^-(dq_1^-)\|/M$$
$$\leq 2\tilde{\varepsilon}_0(1 + C/\sqrt{M}) \|dv_1\|/M$$

Combining all these bounds gives

$$\|dv_1 - dv_1'\| \leq (\varepsilon_0 + 2\tilde{\varepsilon}_0/M)\,(1 + C/\sqrt{M})\,\|dv_1\| + C(\delta_0 + \delta_1)\,\|dv_1\|/\cos\varphi_1$$

Lastly, we consider the vectors dV and dV'. These do not change between collisions or due to collisions of the light particle with $\partial \mathcal{D}$. At an interparticle collision, we have, in a way similar to the previous estimates

$$\|dV_1 - dV_1'\| \leq \|dV_0 - dV_0'\| + 2\|dv_0 - dv_0'\|/M$$
$$+ C\delta_1 \|dv_0\|/M + C\delta_1 \|dq_1^-\|/M$$
$$+ \|\Theta^-\| \|dq_1^- - (dq_1')^-\|/M$$
$$+ \|\Theta^-\| \|dQ_1^- - (dQ_1')^-\|/M$$
$$\text{(C.11)} \qquad + C\delta_1 \|\Theta^-\| \|dq_1^-\|/(M\cos\varphi_1)$$

Applying (C.5) and the same analysis of the operator Θ^- as before gives

$$\|\Theta^-\| \|dq_1^- - (dq_1')^-\|/M \leq \varepsilon_0(1 + C/\sqrt{M}) \|\Theta^-(dq_1^-)\|/M$$
$$+ C(\delta_0 + \delta_1) \|dq_1^-\|/(M\cos\varphi_1)$$

The first term on the right hand side can be combined with the term $2\|dv_0 - dv_0'\|/M$ in (C.11), and we get

$$A'' := 2\|dv_0 - dv_0'\|/M + \varepsilon_0(1 + C/\sqrt{M})\,\|\Theta^-(dq_1^-)\|/M$$
$$\leq 2\varepsilon_0\|dv_0\|/M + 2\varepsilon_0(1 + C/\sqrt{M})\,\|\Theta^-(dq_1^-)\|/M$$
$$\leq 2\varepsilon_0(1 + C/\sqrt{M})\,\|dv_1\|/M$$

We collect the above estimates and obtain

$$\|dV_1 - dV_1'\| \leq (2\varepsilon_0 + \tilde{\varepsilon}_0)\,(1 + C/\sqrt{M})\,\|dv_1\|/M$$
$$+ C(\delta_0 + \delta_1)\,\|dv_1\|/(M\cos\varphi_1)$$

Lemma C.1 is proved. $\square$

Remark. By fixing Q and setting $M = \infty$ we obtain a version of the above lemma for the billiard map $\mathcal{F}_Q$. It becomes much simpler, of course, since $dQ_i = dV_i = 0$ and $\tilde{\varepsilon}_i = 0$, so (C.3) reduces to

$$\varepsilon_1 = \varepsilon_0 + C\frac{\delta_0 + \delta_1}{\cos\varphi_1}$$

and (C.4) becomes obsolete. It is important to note also that the constant C is uniform over all $\mathbf{r} > 0$, as long as the points x_i and x_i' belong to the same u-curve or s-curve (see Section 4.2). To verify the uniformity of C, assume that $q_1 \in \partial\mathcal{P}(Q)$, then $|r_1 - r_1'| \leq c\mathbf{r}|\varphi_1 - \varphi_1'|$ for some $c > 0$, hence $|n(q_1) - n(q_2)| \leq \mathbf{r}^{-1}|r_1 - r_1'| \leq c|\varphi_1 - \varphi_1'| < c\delta_1$, which allows us to suppress the large factor $\mathbf{r}^{-1}$. Hence, the resulting distortion and curvature bounds will be uniform over all $\mathbf{r} > 0$.

COROLLARY C.2. *Suppose that the point x_1 in Lemma C.1 belongs to an H-curve W_1. In that case the key estimates (C.3)–(C.4) of Lemma C.1 can be modified as follows:*

$$(\text{C.12}) \qquad \varepsilon_1 = \left(\varepsilon_0 + \frac{2\tilde{\varepsilon}_0}{M}\right)\left(1 + \frac{C}{\sqrt{M}}\right) + C\frac{\delta_0 + \delta_1}{|W_1|^{2/3}}$$

$$(\text{C.13}) \qquad \tilde{\varepsilon}_1 = \left(2\varepsilon_0 + \tilde{\varepsilon}_0\right)\left(1 + \frac{C}{\sqrt{M}}\right) + C\frac{\delta_0 + \delta_1}{|W_1|^{2/3}}$$

with some constant $C > 1$ (possibly different from the constant in Lemma C.1).

Proof. Indeed, if the points x_1 and x_1' lie in a homogeneity section $\mathbb{H}_k$, then $|W_1| \leq \text{Const}\,(\cos\varphi_1)^{3/2}$, see (4.17). This proves Corollary C.2. $\square$

C. DISTORTION BOUNDS

We now extend the estimates of Lemma C.1 and Corollary C.2 to an arbitrary iteration of $\mathcal{F}$. We will show that the tangent vectors dx_n and dx'_n are $(\varepsilon_n, \tilde{\varepsilon}_n)$-close with some ε_n and $\tilde{\varepsilon}_n$ that we will estimate. For brevity, put $\boldsymbol{\varepsilon}_i = (\varepsilon_i, \tilde{\varepsilon}_i)^T$ for $i \leq n$. The bounds in Lemma C.1 and Corollary C.2 can be rewritten in a matrix form

(C.14) $$\boldsymbol{\varepsilon}_i = \mathbf{A}\boldsymbol{\varepsilon}_{i-1} + \mathbf{b}_i$$

where $\mathbf{A}$ is a fixed matrix

$$\mathbf{A} = (1 + C/\sqrt{M})\,\mathbf{B}, \qquad \mathbf{B} = \begin{pmatrix} 1 & 2/M \\ 2 & 1 \end{pmatrix}$$

and $\mathbf{b}_i = (b_i, b_i)^T$, where $b_i = C(\delta_{i-1} + \delta_i)/\cos\varphi_i$ or $b_i = C(\delta_{i-1} + \delta_i)/|W_i|^{2/3}$ depending on whether we are applying (C.3)–(C.4) or (C.12)–(C.13).

Iterating (C.14) gives

(C.15) $$\boldsymbol{\varepsilon}_n = \mathbf{A}^n \boldsymbol{\varepsilon}_0 + \sum_{i=1}^n \mathbf{A}^{n-i}\mathbf{b}_i$$

The matrix $\mathbf{B}$ has eigenvalues $\lambda_1 = 1 + 2/\sqrt{M}$ and $\lambda_2 = 1 - 2/\sqrt{M}$. By using its eigenvectors, we find

$$\mathbf{B}^k = \tfrac{1}{2} \begin{pmatrix} \lambda_1^k + \lambda_2^k & (\lambda_1^k - \lambda_2^k)/\sqrt{M} \\ (\lambda_1^k - \lambda_2^k)\sqrt{M} & \lambda_1^k + \lambda_2^k \end{pmatrix}$$

It is easy to see that $\|B^k\| \leq \text{Const}\, k\, (1 + 2/\sqrt{M})^k$, thus

$$\|A^k\| \leq C_1\, k\, \left(1 + \frac{C_2}{\sqrt{M}}\right)^k.$$

For some constants $C_1, C_2 > 0$. This gives us

LEMMA C.3 (n step distortion control). *Assume that the standard unstable vectors dx_0 and dx'_0 are $(\varepsilon_0, \varepsilon_0)$-close for some small $\varepsilon_0 > 0$. Then their images dx_n and dx'_n are $(\varepsilon_n, \varepsilon_n)$-close with*

$$\varepsilon_n = C_1 n \left(1 + \frac{C_2}{\sqrt{M}}\right)^n \varepsilon_0 + C_1 \sum_{i=1}^n (n-i+1)\left(1 + \frac{C_2}{\sqrt{M}}\right)^{n-i+1} b_i$$

for all $n \leq 1$.

We see that the sequence ε_n effectively grows linearly with n.

Now we are ready to prove Propositions 4.6 and 4.7. For brevity, we will say ε-close instead of $(\varepsilon, \varepsilon)$-close.

LEMMA C.4. *Under the assumptions of Proposition 4.6, for any $c > 0$ there is a $C > 0$ such that whenever the tangent vectors dx_0 and dx'_0 are ε_0-close with $\varepsilon_0 = c|W_0(x_0, x'_0)|/|W_0|^{2/3}$, then the tangent vectors dx_i and dx'_i are ε_i-close with $\varepsilon_i = C|W_i(x_i, x'_i)|/|W_i|^{2/3}$ for all $i = 1, \ldots, n$.*

Proof. Since the points x_i and x'_i belong to one H-curve, we can redefine δ_i to be $|W_i(x_i, x'_i)|$, and all our previous estimates will hold (with maybe different values of the constants). Next, since $\delta_0 \simeq |W_0|$, the initial tangent vectors dx_0 and dx'_0 are $(c\delta_0^{1/3})$-close. Similarly, we have $\delta_{i-1} + \delta_i \leq \text{Const}\,\delta_i$, hence $b_i \leq \text{Const}\,\delta_i^{1/3}$ in the notation of Lemma C.3. Since H-curves grow by a factor $\vartheta^{-1} > 1$, cf. (4.7), we have $\delta_i \leq \vartheta^{n-i}\delta_n$ for all $i < n$. We now employ Lemma C.3 and easily obtain that the tangent vectors dx_n and dx'_n are $(C\delta_n^{1/3})$-close with some $C > 0$. Therefore

$$\left|\ln \frac{\mathcal{J}_{W_0}\mathcal{F}^n(x_0)}{\mathcal{J}_{W_0}\mathcal{F}^n(x'_0)}\right| \leq C\delta_n^{1/3} \tag{C.16}$$

for some $C > 0$. This estimate is weaker than the distortion bound claimed in Proposition 4.6, but it provides us, at least, with a uniform bound on distortions in the sense of (4.18) with some $\tilde{\beta} > 0$.

The exponential growth of H-curves (4.7) and the uniform bound (4.18) imply that

$$\frac{\delta_i}{|W_i|^{2/3}} \leq C\vartheta^{\frac{n-i}{3}} \frac{\delta_n}{|W_n|^{2/3}}$$

for all $i < n$ and some constant $C > 0$. Now we apply (C.14)–(C.15) with $b_i = \text{Const}\,\delta_i/|W_i|^{2/3}$ and easily obtain that the tangent vectors dx_n and dx'_n are $(C\delta_n/|W_n|^{2/3})$-close with some $C > 0$. Lemma C.4 is proved. □

Now Propositions 4.6 and 4.7 follow directly. □

Remark. It is clear that for a sufficiently smooth unstable curve one can always choose tangent vectors at any two points that are ε-close for arbitrarily small $\varepsilon > 0$, thus they will satisfy the assumptions of Lemma C.4.

Lastly, we prove Lemma 4.20. Let dy' and dy'' be tangent vectors to the curve γ at the points y' and y'', respectively. According to the definition of standard pairs, we can assume that they are ε-close with $\varepsilon = C\,\text{dist}(y', y'')/|\gamma|^{2/3}$. Then by Proposition 4.6, which we just

proved, the vectors $dx'_- = D\mathcal{F}^i(dy')$ and $dx''_- = D\mathcal{F}^i(dy'')$ are ε_--close with
$$\varepsilon_- = C\,\mathrm{dist}(x'_-, x''_-)/|W|^{2/3} \leq \mathrm{Const}\,\varepsilon_\gamma/|W|^{2/3}$$
We now compare the tangent vectors $dx' = (D\mathcal{F}_Q \circ D\pi_0)(dx'_-)$ and $dx'' = (D\pi_0 \circ D\mathcal{F})(dx''_-)$.

Claim. dx' and dx'' are ε_0-close with

(C.17) $$\varepsilon_0 = \frac{C\varepsilon_\gamma}{|W|^{2/3}} + \frac{C\varepsilon_\gamma}{|W'_0|^{2/3}}$$

where $C > 0$ is a large constant.

Proof. Our argument follows the same lines as the proofs of Lemma C.1 and Corollary C.2, and we only focus on the novelty of the present situation. First, since $\mathrm{dist}(x'_-, x''_-) = \mathcal{O}(\varepsilon_\gamma)$ and $\mathrm{dist}(x', x'') = \mathcal{O}(\varepsilon_\gamma)$, then both δ_0 and δ_1 in (C.12)–(C.13) will be $\mathcal{O}(\varepsilon_\gamma)$. In addition, we apply $D\pi_0$ to both vectors. Recall that the projection $\pi_0\colon \Omega \to \Omega_0$, fixes the position of the heavy particle, sets its velocity to zero, and normalizes the vector w defined by (1.5). Accordingly, $D\pi_0$ sets the components dQ and dV of the tangent vector to zero and rescales the component dw by the same factor as it rescales w, i.e. it divides dw by $\|w\|$. In addition, we need to project both components dq and dw onto the line perpendicular to w, so that the basic equations (4.5)–(4.6) would hold. Therefore, the map $D\pi_Q\colon (dQ, dV, dq, dw) \mapsto (dQ_1, dV_1, dq_1, dw_1)$ acts according to the following rules: $dQ_1 = dV_1 = 0$, and
$$dw_1 = dw/\|w\| - \langle dw, w\rangle/\|w\|^3, \qquad dq_1 = dq - \langle dq, w\rangle/\|w\|^2,$$
As we noted in Section 4.1, the estimates (a)–(g) of Proposition 4.1 apply to the vectors w and dw, just as well as to v and dv. Hence
$$\|dq_1 - dq\| \leq C\|V\|\,\|dq\| \leq C\varepsilon_\gamma\|dq\|$$
and
$$\big\|\,\|w\|\,dw_1 - dw\,\big\| \leq C\|V\|\,\|dw\| \leq C\varepsilon_\gamma\|dw\|$$
Such a difference can be incorporated into the right hand side of (C.17). The division of dw by $\|w\|$ results in a change of order one, in general, but this will be matched by the corresponding division by $\|w\|$ when $D\pi_0$ is applied to the other vector, as one can easily verify. This completes the proof of the claim. $\square$

We now finish the proof of Lemma 4.20. For each $r \geq 1$ we need to compare the tangent vectors $dx'_r = D\mathcal{F}^r_Q(dx')$ and $dx''_r = D\mathcal{F}^r_Q(dx'')$. The map $\mathcal{F}_Q$ on Ω_Q corresponds to the motion of the light particle when the heavy one is fixed at Q, which is the limit case of our two-particle dynamics as $M \to \infty$. Thus, our analysis in Appendix B, in particular Lemma C.1, Corollary C.2, and Lemma C.3 apply to the map $\mathcal{F}_Q$ as well. In order to use them, though, we need to verify the conditions they are based on. First, since for each $r \geq 1$ the points $\mathcal{F}_Q^r(x')$ and $\mathcal{F}_Q^r(x'')$ belong to one homogeneous stable manifold, they lie in one homogeneity section. Second, (C.1)–(C.2) hold due to (4.14). Now Corollary C.2 and Lemma C.3 can be used, indeed, and they directly imply Lemma 4.20. □

Bibliography

[1] Bakhtin V. I. *A direct method for constructing an invariant measure on a hyperbolic attractor*, Russian Acad. Sci. Izv. Math. **41** (1993), 207–227.

[2] Bakhtin V. I. *On the averaging method in a system with fast hyperbolic motions*, Proc. Belorussian Math. Inst. **6** (2000), 23–26.

[3] Bernstein S. N. *Sur l'extension du theoreme limite de calcul des probabilites aux sommes de quantites dependentes*, Math. Ann. **97** (1926), 1–59.

[4] Bleher P. M. *Statistical properties of two-dimensional periodic Lorentz gas with infinite horizon*, J. Stat. Phys. **66** (1992), 315–373.

[5] Bonetto F., Daems D., and Lebowitz J. *Properties of stationary nonequilibrium states in the thermostated periodic Lorentz gas I: the one particle system*, J. Stat. Phys. **101** (2000), 35–60.

[6] Bonetto F., Lebowitz J., and Rey-Bellet L. *Fourier Law: a challenge to theorists*, in Mathematical Physics-2000, Imperial College Press, London, 2000.

[7] Bressaud X. and Liverani C. *Anosov diffeomorphism and coupling*, Ergod. Th. Dynam. Syst., **22** (2002), 129–152.

[8] Bunimovich L. A. *On the ergodic properties of nowhere dispersing billiards*, Comm. Math. Phys. **65** (1979), 295–312.

[9] Bunimovich L. A. and Sinai Ya. G. *Statistical properties of Lorentz gas with periodic configuration of scatterers*, Comm. Math. Phys. **78** (1980/81), 479–497.

[10] Bunimovich L. A., Sinai Ya. G., and Chernov N. I. *Markov partitions for two-dimensional hyperbolic billiards*, Russ. Math. Surv. **45** (1990), 105–152.

[11] Bunimovich L. A., Sinai Ya. G., and Chernov N. I. *Statistical properties of two-dimensional hyperbolic billiards*, Russ. Math. Surv. **46** (1991), 47–106.

[12] Bunimovich L., Liverani C., Pellegrinotti A., Suhov Y. *Ergodic systems of n balls in a billiard table*, Comm. Math. Phys. **146** (1992), 357–396.

[13] Bunimovich, L. A. and Spohn, H. *Viscosity for a periodic two disk fluid: an existence proof*, Comm. Math. Phys. **176** (1996), 661–680.

[14] Calderoni P., Durr D., and Kusuoka S. *A mechanical model of Brownian motion in half-space*, J. Stat. Phys. **55** (1989), 649–693.

[15] Chernov N. *Statistical properties of the periodic Lorentz gas. Multidimensional case*, J. Stat. Phys. **74** (1994), 11–53.

[16] Chernov N. *Entropy, Lyapunov exponents and mean-free path for billiards*, J. Stat. Phys. **88** (1997), 1–29.

[17] Chernov N. I. *Markov approximations and decay of correlations for Anosov flows*, Ann. Math. **147** (1998), 269–324.

[18] Chernov N. *Decay of correlations in dispersing billiards*, J. Stat. Phys. **94** (1999), 513–556.

[19] Chernov N. *Sinai billiards under small external forces*, Ann. H. Poincaré **2** (2001), 197–236.

[20] Chernov N. *Advanced statistical properties of dispersing billiards*, J. Stat. Phys. **122** (2006), 1061–1094.

[21] Chernov N. and Dettmann C. P. *The existence of Burnett coefficients in the periodic Lorentz gas*, Phys. A **279** (2000), 37–44.

[22] Chernov N. and Dolgopyat D. *Hyperbolic billiards and statistical physics*, in Proc. ICM (Madrid, Spain, 2006), Vol II, Euro. Math. Soc., Zurich, 2006, pp. 1679-1704.

[23] Chernov N., Eyink G. L., Lebowitz J. L., and Sinai Ya. G. *Steady-state electrical conduction in the periodic Lorentz gas*, Comm. Math. Phys. **154** (1993), 569–601.

[24] Chernov N., Lebowitz J., and Sinai Ya. *Scaling dynamic of a massive piston in a cube filled with ideal gas: Exact results*, J. Stat. Phys. **109** (2002), 529–548.

[25] Chernov N., Lebowitz J., and Sinai Ya. *Dynamic of a massive piston in an ideal gas*, Russ. Math. Surv. **57** (2002), 1–84.

[26] Chernov N. and Lebowitz J. *Dynamics of a massive piston in an ideal gas: Oscillatory motion and approach to equilibrium*, J. Stat. Phys. **109** (2002), 507–527.

[27] Chernov N. and Markarian R. *Dispersing billiards with cusps: slow decay of correlations*, Comm. Math. Phys. **270** (2007), 727–758.

[28] Chernov N. and Markarian R. *Chaotic billiards*, Math. Surveys and Monographs, **127** AMS, Providence, RI, 2006.

[29] Chernov N., Markarian R., and Troubetzkoy S. *Invariant measures for Anosov maps with small holes*, Erg. Th. Dyn. Sys. **20** (2000), 1007–1044.

[30] Chernov N. and Zhang H.-K. *Billiards with polynomial mixing rates*, Nonlinearity **18** (2005), 1527–1553.

[31] Cogburn R. and Ellison J. A. *A four-thirds law for phase randomization of stochastically perturbed oscillators and related phenomena*, Comm. Math. Phys. **166** (1994), 317–336.

[32] Conze, J.-P. and Le Borgne, S. *Methode de martingales et flot geodesique sur une surface de courbure constante négative*, Erg. Th. Dyn. Sys. **21** (2001), 421–441.

[33] Denker M. and Philipp W. *Approximation by Brownian motion for Gibbs measures and flows under a function*, Erg. Th. Dyn. Sys. **4** (1984), 541–552.

[34] Dolgopyat D. *On differentiability of SRB states for partially hyperbolic systems*, Invent. Math. **155** (2004), 389–449.

[35] Dolgopyat D. *Limit Theorems for partially hyperbolic systems*, Trans. AMS. **356** (2004), 1637–1689.

[36] Durr D., Goldstein S., and Lebowitz J. L. *A mechanical model of Brownian motion*, Comm. Math. Phys. **78** (1980/81), 507–530.

[37] Durr D., Goldstein S., and Lebowitz J. L. *A mechanical model for the Brownian motion of a convex body*, Z. Wahrsch. Verw. Gebiete **62** (1983), 427–448.

[38] Efimov K., *A Livshits-type theorem for scattering billiards*, Theor. Math. Phys. **98** (1994), 122–131.

[39] Einstein A. *Investigations on the theory of the Brownian movement*, Edited with notes by R. Frth. Dover, New York, 1956.

[40] Field M., Melbourne I., and Török A. *Decay of correlations, central limit theorems and approximation by Brownian motion for compact Lie group extensions,* Erg. Th. Dyn. Syst. **23** (2003), 87-110.

[41] Freidlin M. I. and Wentzell A. D. *Random perturbations of dynamical systems,* Grundlehren der Mathematischen Wissenschaften **260** (1984) Springer-Verlag, New York.

[42] Gaspard P. and Klages R. *Chaotic and fractal properties of deterministic diffusion-reaction processes,* Chaos **8** (1998), 409–423.

[43] Gilbert T. and Dorfman J. R., *On the parametric dependences of a class of non-linear singular maps,* Discr. Cont. Dynam. Syst. Ser. B **4** (2004), 391–406.

[44] Groeneveld J. and Klages R. *Negative and nonlinear response in an exactly solved dynamical model of particle transport,* J. Stat. Phys. **109** (2002), 821–861.

[45] Harayama T. and Gaspard P. *Diffusion of particles bouncing on a one-dimensional periodically corrugated floor,* Phys. Rev. E **64** (2001), 036215.

[46] Harayama T., Klages R., and Gaspard P. *Deterministic diffusion in flower-shaped billiards,* Phys. Rev. E **66** (2002), 026211-1-7.

[47] *Hard Ball Systems and the Lorentz Gas,* Edited by D. Szász. Encyclopaedia of Mathematical Sciences, **101** Springer-Verlag, Berlin, 2000.

[48] Harris T. E. *Diffusion with collisions between particles,* J. Appl. Prob. **2** (1965), 323–338.

[49] Hartman P. *Ordinary differential equations,* 2d ed., Birkhauser, Boston-Basel-Stuttgart, 1982.

[50] Holly R. *The motion of a heavy particle in an infinite one dimensional gas of hard spheres,* Z. Wahrsch. Verw. Gebiete **17** (1971), 181–219.

[51] Ibragimov I. A. and Linnik Yu. V. *Independent and stationary sequences of random variables,* Wolters-Noordhoff Publishing, Groningen, 1971.

[52] Katok A., Knieper G., Pollicott M., and Weiss H. *Differentiability and analyticity of topological entropy for Anosov and geodesic flows,* Invent. Math. **98** (1989), 581–597.

[53] Khasminskii R. Z. *Diffusion processes with a small parameter,* Izv. Akad. Nauk SSSR Ser. Mat. **27** (1963), 1281–1300.

[54] Kifer Yu. *The exit problem for small random perturbations of dynamical systems with a hyperbolic fixed point,* Israel J. Math. **40** (1981), 74–96.

[55] Kifer Yu. *Limit theorems in averaging for dynamical systems,* Erg. Th. Dyn. Sys. **15** (1995), 1143–1172.

[56] Kifer Yu. *Averaging principle for fully coupled dynamical systems and large deviations,* Erg. Th. Dyn. Sys. **24** (2004), 847–871.

[57] Klages R. *Deterministic diffusion in one-dimensional chaotic dynamical systems,* Wissenschaft & Technik Veralg, Berlin, 1996.

[58] Klages R. and Dellago C. *Density-dependent diffusion in the periodic Lorentz gas,* J. Statist. Phys. **101** (2000), 145–159.

[59] Klages R. and Dorfman J. R., *Simple maps with fractal diffusion coefficients,* Phys. Rev. Lett. **74** (1995), 387–390.

[60] Klages R. and Dorfman J. R., *Simple deterministic dynamical systems with fractal diffusion coefficients,* Phys. Rev. E **59** (1999), 5361–5383.

[61] Klages R. and Korabel N. *Understanding deterministic diffusion by correlated random walks,* J. Phys. A **35** (2002), 4823–4836.

[62] Korabel N. and Klages R. *Fractal structures of normal and anomalous diffusion in nonlinear nonhyperbolic dynamical systems*, Phys. Rev. Lett. **89** (2002), 214102-1-4.

[63] Krámli A., Simányi N., Szász D. *A "transversal" fundamental theorem for semi-dispersing billiards*, Comm. Math. Phys. **129** (1990), 535–560.

[64] Lazutkin V. F. *Existence of caustics for the billiard problem in a convex domain*, (Russian) Izv. Akad. Nauk SSSR Ser. Mat. **37** (1973), 186–216.

[65] Liverani C. *On contact Anosov flows*, Ann. Math. **159** (2004), 1275–1312.

[66] Liverani C. and Wojtkowski M. *Ergodicity in Hamiltonian systems*, in: Dynamics reported, 130–202, Dynam. Report. Expositions Dynam. Systems (N.S.), **4**, Springer, Berlin, 1995.

[67] Machta J. *Power Law Decay of Correlations in a Billiard Problem*, J. Stat. Phys. **32** (1983), 555–564.

[68] Major P. and Szasz D. *On the effect of collisions on the motion of an atom in $\mathbb{R}^1$*, Ann. Prob. **8** (1980), 1068–1078.

[69] Melbourne I. and Török A. *Statistical limit theorems for suspension flows*, Israel J. Math. **144** (2004), 191–209.

[70] Nelson E. *Dynamical theories of Brownian motion*, Princeton University Press, Princeton, N.J. 1967.

[71] Pardoux E. and Veretennikov A. Yu. *On the Poisson equation and diffusion approximation-I*, Ann. Probab. **29** (2001), 1061–1085.

[72] Parry W. and Pollicott M. *Zeta functions and the periodic orbit structure of hyperbolic dynamics*, Asterisque **187-188** (1990).

[73] Pène F. *Averaging method for differential equations perturbed by dynamical systems*, ESAIM Probab. Stat. **6** (2002), 33–88 (electronic).

[74] Pesin Ya. B. and Sinai Ya. G. *Gibbs measures for partially hyperbolic attractors*, Erg. Th. Dyn. Sys. **2** (1982), 417–438.

[75] Ratner M. *The central limit theorem for geodesic flows on n-dimensional manifolds of negative curvature*, Israel J. Math. **16** (1973), 181–197

[76] Revuz D. and Yor M. *Continuous Martingales and Brownian Motion*, 3d ed., Sringer, Berlin-Heildelberg-New York, 1998.

[77] Ruelle D. *Invariant measures for a diffeomorphism which expands the leaves of a foliation*, Publ. IHES. **48** (1978), 133–135.

[78] Ruelle D. *Differentiation of SRB states*, Comm. Math. Phys. **187** (1997), 227–241.

[79] Ruelle D. *Differentiation of SRB states: Corrections and complements*, Comm. Math. Phys. **234** (2003), 185–190.

[80] Shub M. *Global stability of dynamical systems*, With the collaboration of Albert Fathi and Remi Langevin, Springer-Verlag, New York, 1987.

[81] Simányi N. *Ergodicity of hard spheres in a box*, Ergod. Th. Dynam. Syst. **19** (1999), 741–766.

[82] Simányi N. *Proof of the Boltzmann-Sinai ergodic hypothesis for typical hard disk systems*, Invent. Math., **154** (2003), 123–178.

[83] Simányi N. *Proof of the ergodic hypothesis for typical hard ball systems*, Ann. H. Poincaré **5** (2004), 203–233.

[84] Simányi N. and Szász D. *Hard ball systems are completely hyperbolic*, Ann. Math. **149** (1999), 35–96.

[85] Sinai Ya. G. *Classical dynamic systems with countably-multiple Lebesgue spectrum: I,II,* Izv. Akad. Nauk SSSR Ser. Mat. **25** (1961), 899–924, **30** (1966), 15–68.

[86] Sinai Ya. G. *Dynamical systems with elastic reflections: Ergodic properties of dispersing billiards,* Russ. Math. Surv. **25** (1970), 137–189.

[87] Sinai Ya. G. and Soloveichik M. *One dimensional classical particle in the ideal gas,* Comm. Math. Phys. **104** (1986), 424–443.

[88] Soloveitchik M. *Mechanical background of Brownian motion,* in Sinai's Moscow Seminar on Dynamical Systems, AMS Transl. Ser. 2, **171** AMS, Providence, RI, 1996, 233–247.

[89] Spitzer F. *Uniform motion with elastic collision of an infinite particle systems,* J. Math. Mech. **18** (1969), 973–989.

[90] Spohn H. *Large scale dynamics of interacting particles,* Springer-Verlag, Berlin, New York, 1991.

[91] Stroock D. W. and Varadhan S. R. S. *Multidimensional diffusion processes,* Grundlehren der Mathematischen Wissenschaften **233,** (1979) Springer-Verlag, Berlin-New York.

[92] Szász D. and Toth B. *Bounds for the limiting variance of the "heavy particle" in R^1,* Comm. Math. Phys. **104** (1986), 445–455.

[93] Szász D. and Toth B. *Towards a unified dynamical theory of the Brownian particle in an ideal gas,* Comm. Math. Phys. **111** (1987), 41–62.

[94] Szatzschneider W. *A more detrministic version of Harris-Spitzer random constant velocity model for infinite systems of particles,* in Lect. Notes Math. **472** (1975) 157–167, Springer.

[95] Varadhan S. R. S. *Regularity of self-diffusion coefficient,* in *The Dynkin Festschrift* (Ed. M. I. Freidlin) Progr. Prob., Birkhauser Boston **34** (1994), 387–397.

[96] Young L.-S. *Statistical properties of dynamical systems with some hyperbolicity,* Ann. Math. **147** (1998), 585–650.

[97] Young L.-S. *Recurrence times and rates of mixing,* Israel J. Math. **110** (1999), 153–188.

Index

Auxiliary measures, 15–18, 20, 95, 96
Averaging, 12, 14

Big-small block technique, 99, 155

Coupling, 140–142, 148

Diffusion matrix, 10, 16, 17, 57, 62, 133, 134, 158, 159, 162
Dispersing billiards, 2–5, 15, 48, 58, 62, 91, 133, 139, 150, 167
Distortion bounds, 34, 35, 39, 41, 88, 89, 151, 177, 182, 184

Einstein relation, 61
Equidistribution, 15, 16, 21, 22, 25, 47, 48, 99, 140

Finite horizon, 2, 3, 13, 18, 38, 44, 46, 58, 139, 149

Green-Kubo formula, 5, 158
Growth lemma, 37, 76, 89, 91, 151, 153, 158, 161, 165

H-components, 35–41, 47–53, 55, 71, 76, 85, 88–90, 140, 141, 143–146, 148, 151–154, 156, 157, 167
H-curves, 35–37, 39, 41, 43, 44, 47, 49–51, 54, 71, 140–143, 146, 151, 152, 161, 167, 182, 184
Holonomy map, 48, 50, 51, 53, 145, 146
Homogeneity sections, 33, 35, 38, 177, 182, 186
Homogeneity strips, 33, 43, 50, 71, 85, 88, 151, 154

Log-Lipschitz continuity, 10, 16, 17, 57, 62

Ohm's law, 60

Rayleigh gas, 61

Scatterer, 1, 3, 10, 28, 31, 37, 38, 58, 59, 61, 136, 139, 149–152, 159, 162
Shadowing, 15, 48, 51, 53, 63, 140
Standard pair, 17, 18, 20, 21, 25, 27, 36, 37, 41, 47, 75, 82, 94–96, 100, 103, 104, 109, 110, 113, 116, 117, 124–126, 130, 140–144, 147, 148, 155, 156, 161, 184

u-curves (unstable curves), 29–35, 42–44, 47, 48, 50, 66, 76, 79, 82, 90, 152, 154, 161, 164, 167, 168, 177, 182
Unstable vectors, 27–29, 33, 65, 177, 178, 183

Editorial Information

To be published in the *Memoirs*, a paper must be correct, new, nontrivial, and significant. Further, it must be well written and of interest to a substantial number of mathematicians. Piecemeal results, such as an inconclusive step toward an unproved major theorem or a minor variation on a known result, are in general not acceptable for publication.

Papers appearing in *Memoirs* are generally at least 80 and not more than 200 published pages in length. Papers less than 80 or more than 200 published pages require the approval of the Managing Editor of the Transactions/Memoirs Editorial Board.

As of November 30, 2008, the backlog for this journal was approximately 11 volumes. This estimate is the result of dividing the number of manuscripts for this journal in the Providence office that have not yet gone to the printer on the above date by the average number of monographs per volume over the previous twelve months, reduced by the number of volumes published in four months (the time necessary for preparing a volume for the printer). (There are 6 volumes per year, each usually containing at least 4 numbers.)

A Consent to Publish and Copyright Agreement is required before a paper will be published in the *Memoirs*. After a paper is accepted for publication, the Providence office will send a Consent to Publish and Copyright Agreement to all authors of the paper. By submitting a paper to the *Memoirs*, authors certify that the results have not been submitted to nor are they under consideration for publication by another journal, conference proceedings, or similar publication.

Information for Authors

Memoirs are printed from camera copy fully prepared by the author. This means that the finished book will look exactly like the copy submitted.

Initial submission. The AMS uses Centralized Manuscript Processing for initial submissions. Authors should submit a PDF file using the Initial Manuscript Submission form found at www.ams.org/peer-review-submission, or send one copy of the manuscript to the following address: Centralized Manuscript Processing, MEMOIRS OF THE AMS, 201 Charles Street, Providence, RI 02904-2294 USA. If a paper copy is being forwarded to the AMS, indicate that it is for it Memoirs and include the name of the corresponding author, contact information such as email address or mailing address, and the name of an appropriate Editor to review the paper (see the list of Editors below).

The paper must contain a *descriptive title* and an *abstract* that summarizes the article in language suitable for workers in the general field (algebra, analysis, etc.). The *descriptive title* should be short, but informative; useless or vague phrases such as "some remarks about" or "concerning" should be avoided. The *abstract* should be at least one complete sentence, and at most 300 words. Included with the footnotes to the paper should be the 2000 *Mathematics Subject Classification* representing the primary and secondary subjects of the article. The classifications are accessible from www.ams.org/msc/. The list of classifications is also available in print starting with the 1999 annual index of *Mathematical Reviews*. The Mathematics Subject Classification footnote may be followed by a list of *key words and phrases* describing the subject matter of the article and taken from it. Journal abbreviations used in bibliographies are listed in the latest *Mathematical Reviews* annual index. The series abbreviations are also accessible from www.ams.org/msnhtml/serials.pdf. To help in preparing and verifying references, the AMS offers MR Lookup, a Reference Tool for Linking, at www.ams.org/mrlookup/.

Electronically prepared manuscripts. The AMS encourages electronically prepared manuscripts, with a strong preference for $\mathcal{A}_{\mathcal{M}}\mathcal{S}$-LaTeX. To this end, the Society has prepared $\mathcal{A}_{\mathcal{M}}\mathcal{S}$-LaTeX author packages for each AMS publication. Author packages include instructions for preparing electronic manuscripts, samples, and a style file that generates the particular design specifications of that publication series. Though $\mathcal{AMS}$-LaTeX is the highly preferred format of TeX, author packages are also available in $\mathcal{AMS}$-TeX.

Authors may retrieve an author package for *Memoirs of the AMS* from www.ams.org/journals/memo/memoauthorpac.html or via FTP to ftp.ams.org (login as anonymous, enter username as password, and type cd pub/author-info). The *AMS Author Handbook* and the *Instruction Manual* are available in PDF format from the author package link. The author package can also be obtained free of charge by sending email to tech-support@ams.org (Internet) or from the Publication Division, American Mathematical Society, 201 Charles St., Providence, RI 02904-2294, USA. When requesting an author package, please specify $\mathcal{AMS}$-LaTeX or $\mathcal{AMS}$-TeX and the publication in which your paper will appear. Please be sure to include your complete mailing address.

After acceptance. The final version of the electronic file should be sent to the Providence office (this includes any TeX source file, any graphics files, and the DVI or PostScript file) immediately after the paper has been accepted for publication.

Before sending the source file, be sure you have proofread your paper carefully. The files you send must be the EXACT files used to generate the proof copy that was accepted for publication. For all publications, authors are required to send a printed copy of their paper, which exactly matches the copy approved for publication, along with any graphics that will appear in the paper.

Accepted electronically prepared files can be submitted via the web at www.ams.org/submit-book-journal/, sent via FTP, or sent on CD-Rom or diskette to the Electronic Prepress Department, American Mathematical Society, 201 Charles Street, Providence, RI 02904-2294 USA. TeX source files, DVI files, and PostScript files can be transferred over the Internet by FTP to the Internet node ftp.ams.org (130.44.1.100). When sending a manuscript electronically via CD-Rom or diskette, please be sure to include a message identifying the paper as a Memoir.

Electronically prepared manuscripts can also be sent via email to pub-submit@ams.org (Internet). In order to send files via email, they must be encoded properly. (DVI files are binary and PostScript files tend to be very large.)

Electronic graphics. Comprehensive instructions on preparing graphics are available at www.ams.org/authors/journals.html. A few of the major requirements are given here.

Submit files for graphics as EPS (Encapsulated PostScript) files. This includes graphics originated via a graphics application as well as scanned photographs or other computer-generated images. If this is not possible, TIFF files are acceptable as long as they can be opened in Adobe Photoshop or Illustrator. No matter what method was used to produce the graphic, it is necessary to provide a paper copy to the AMS.

Authors using graphics packages for the creation of electronic art should also avoid the use of any lines thinner than 0.5 points in width. Many graphics packages allow the user to specify a "hairline" for a very thin line. Hairlines often look acceptable when proofed on a typical laser printer. However, when produced on a high-resolution laser imagesetter, hairlines become nearly invisible and will be lost entirely in the final printing process.

Screens should be set to values between 15% and 85%. Screens which fall outside of this range are too light or too dark to print correctly. Variations of screens within a graphic should be no less than 10%.

Inquiries. Any inquiries concerning a paper that has been accepted for publication should be sent to memo-query@ams.org or directly to the Electronic Prepress Department, American Mathematical Society, 201 Charles St., Providence, RI 02904-2294 USA.

Editors

This journal is designed particularly for long research papers, normally at least 80 pages in length, and groups of cognate papers in pure and applied mathematics. Papers intended for publication in the *Memoirs* should be addressed to one of the following editors. The AMS uses Centralized Manuscript Processing for initial submissions to AMS journals. Authors should follow instructions listed on the Initial Submission page found at www.ams.org/memo/memosubmit.html.

Algebra to ALEXANDER KLESHCHEV, Department of Mathematics, University of Oregon, Eugene, OR 97403-1222; email: ams@noether.uoregon.edu

Algebraic geometry to DAN ABRAMOVICH, Department of Mathematics, Brown University, Box 1917, Providence, RI 02912; email: amsedit@math.brown.edu

Algebraic geometry and its application to MINA TEICHER, Emmy Noether Research Institute for Mathematics, Bar-Ilan University, Ramat-Gan 52900, Israel; email: teicher@macs.biu.ac.il

Algebraic topology to ALEJANDRO ADEM, Department of Mathematics, University of British Columbia, Room 121, 1984 Mathematics Road, Vancouver, British Columbia, Canada V6T 1Z2; email: adem@math.ubc.ca

Combinatorics to JOHN R. STEMBRIDGE, Department of Mathematics, University of Michigan, Ann Arbor, Michigan 48109-1109; email: FRS@umich.edu

Commutative and homological algebra to LUCHEZAR L. AVRAMOV, Department of Mathematics, University of Nebraska, Lincoln, NE 68588-0130; email: avramov@math.unl.edu

Complex analysis and harmonic analysis to ALEXANDER NAGEL, Department of Mathematics, University of Wisconsin, 480 Lincoln Drive, Madison, WI 53706-1313; email: nagel@math.wisc.edu

Differential geometry and global analysis to LISA C. JEFFREY, Department of Mathematics, University of Toronto, 100 St. George St., Toronto, ON Canada M5S 3G3; email: jeffrey@math.toronto.edu

Dynamical systems and ergodic theory and complex anaysis to YUNPING JIANG, Department of Mathematics, CUNY Queens College and Graduate Center, 65-30 Kissena Blvd., Flushing, NY 11367; email: Yunping.Jiang@qc.cuny.edu

Functional analysis and operator algebras to DIMITRI SHLYAKHTENKO, Department of Mathematics, University of California, Los Angeles, CA 90095; email: shlyakht@math.ucla.edu

Geometric analysis to WILLIAM P. MINICOZZI II, Department of Mathematics, Johns Hopkins University, 3400 N. Charles St., Baltimore, MD 21218; email: trans@math.jhu.edu

Geometric analysis to MARK FEIGHN, Math Department, Rutgers University, Newark, NJ 07102; email: feighn@andromeda.rutgers.edu

Harmonic analysis, representation theory, and Lie theory to ROBERT J. STANTON, Department of Mathematics, The Ohio State University, 231 West 18th Avenue, Columbus, OH 43210-1174; email: stanton@math.ohio-state.edu

Logic to STEFFEN LEMPP, Department of Mathematics, University of Wisconsin, 480 Lincoln Drive, Madison, Wisconsin 53706-1388; email: lempp@math.wisc.edu

Number theory to JONATHAN ROGAWSKI, Department of Mathematics, University of California, Los Angeles, CA 90095; email: jonr@math.ucla.edu

Number theory to SHANKAR SEN, Department of Mathematics, 505 Malott Hall, Cornell University, Ithaca, NY 14853; email: ss70@cornell.edu

Partial differential equations to GUSTAVO PONCE, Department of Mathematics, South Hall, Room 6607, University of California, Santa Barbara, CA 93106; email: ponce@math.ucsb.edu

Partial differential equations and dynamical systems to PETER POLACIK, School of Mathematics, University of Minnesota, Minneapolis, MN 55455; email: polacik@math.umn.edu

Probability and statistics to RICHARD BASS, Department of Mathematics, University of Connecticut, Storrs, CT 06269-3009; email: bass@math.uconn.edu

Real analysis and partial differential equations to DANIEL TATARU, Department of Mathematics, University of California, Berkeley, Berkeley, CA 94720; email: tataru@math.berkeley.edu

All other communications to the editors should be addressed to the Managing Editor, ROBERT GURALNICK, Department of Mathematics, University of Southern California, Los Angeles, CA 90089-1113; email: guralnic@math.usc.edu.

Titles in This Series

929 **Richard F. Bass, Xia Chen, and Jay Rosen,** Moderate deviations for the range of planar random walks, 2009

928 **Ulrich Bunke,** Index theory, eta forms, and Deligne cohomology, 2009

927 **N. Chernov and D. Dolgopyat,** Brownian Brownian motion-I, 2009

926 **Riccardo Benedetti and Francesco Bonsante,** Canonical Wick rotations in 3-dimensional gravity, 2009

925 **Sergey Zelik and Alexander Mielke,** Multi-pulse evolution and space-time chaos in dissipative systems, 2009

924 **Pierre-Emmanuel Caprace,** "Abstract" homomorphisms of split Kac-Moody groups, 2009

923 **Michael Jöllenbeck and Volkmar Welker,** Minimal resolutions via algebraic discrete Morse theory, 2009

922 **Ph. Barbe and W. P. McCormick,** Asymptotic expansions for infinite weighted convolutions of heavy tail distributions and applications, 2009

921 **Thomas Lehmkuhl,** Compactification of the Drinfeld modular surfaces, 2009

920 **Georgia Benkart, Thomas Gregory, and Alexander Premet,** The recognition theorem for graded Lie algebras in prime characteristic, 2009

919 **Roelof W. Bruggeman and Roberto J. Miatello,** Sum formula for SL_2 over a totally real number field, 2009

918 **Jonathan Brundan and Alexander Kleshchev,** Representations of shifted Yangians and finite W-algebras, 2008

917 **Salah-Eldin A. Mohammed, Tusheng Zhang, and Huaizhong Zhao,** The stable manifold theorem for semilinear stochastic evolution equations and stochastic partial differential equations, 2008

916 **Yoshikata Kida,** The mapping class group from the viewpoint of measure equivalence theory, 2008

915 **Sergiu Aizicovici, Nikolaos S. Papageorgiou, and Vasile Staicu,** Degree theory for operators of monotone type and nonlinear elliptic equations with inequality constraints, 2008

914 **E. Shargorodsky and J. F. Toland,** Bernoulli free-boundary problems, 2008

913 **Ethan Akin, Joseph Auslander, and Eli Glasner,** The topological dynamics of Ellis actions, 2008

912 **Igor Chueshov and Irena Lasiecka,** Long-time behavior of second order evolution equations with nonlinear damping, 2008

911 **John Locker,** Eigenvalues and completeness for regular and simply irregular two-point differential operators, 2008

910 **Joel Friedman,** A proof of Alon's second eigenvalue conjecture and related problems, 2008

909 **Cameron McA. Gordon and Ying-Qing Wu,** Toroidal Dehn fillings on hyperbolic 3-manifolds, 2008

908 **J.-L. Waldspurger,** L'endoscopie tordue n'est pas si tordue, 2008

907 **Yuanhua Wang and Fei Xu,** Spinor genera in characteristic 2, 2008

906 **Raphaël S. Ponge,** Heisenberg calculus and spectral theory of hypoelliptic operators on Heisenberg manifolds, 2008

905 **Dominic Verity,** Complicial sets characterising the simplicial nerves of strict ω-categories, 2008

904 **William M. Goldman and Eugene Z. Xia,** Rank one Higgs bundles and representations of fundamental groups of Riemann surfaces, 2008

903 **Gail Letzter,** Invariant differential operators for quantum symmetric spaces, 2008

TITLES IN THIS SERIES

902 **Bertrand Toën and Gabriele Vezzosi,** Homotopical algebraic geometry II: Geometric stacks and applications, 2008
901 **Ron Donagi and Tony Pantev (with an appendix by Dmitry Arinkin),** Torus fibrations, gerbes, and duality, 2008
900 **Wolfgang Bertram,** Differential geometry, Lie groups and symmetric spaces over general base fields and rings, 2008
899 **Piotr Hajłasz, Tadeusz Iwaniec, Jan Malý, and Jani Onninen,** Weakly differentiable mappings between manifolds, 2008
898 **John Rognes,** Galois extensions of structured ring spectra/Stably dualizable groups, 2008
897 **Michael I. Ganzburg,** Limit theorems of polynomial approximation with exponential weights, 2008
896 **Michael Kapovich, Bernhard Leeb, and John J. Millson,** The generalized triangle inequalities in symmetric spaces and buildings with applications to algebra, 2008
895 **Steffen Roch,** Finite sections of band-dominated operators, 2008
894 **Martin Dindoš,** Hardy spaces and potential theory on C^1 domains in Riemannian manifolds, 2008
893 **Tadeusz Iwaniec and Gaven Martin,** The Beltrami Equation, 2008
892 **Jim Agler, John Harland, and Benjamin J. Raphael,** Classical function theory, operator dilation theory, and machine computation on multiply-connected domains, 2008
891 **John H. Hubbard and Peter Papadopol,** Newton's method applied to two quadratic equations in $\mathbb{C}^2$ viewed as a global dynamical system, 2008
890 **Steven Dale Cutkosky,** Toroidalization of dominant morphisms of 3-folds, 2007
889 **Michael Sever,** Distribution solutions of nonlinear systems of conservation laws, 2007
888 **Roger Chalkley,** Basic global relative invariants for nonlinear differential equations, 2007
887 **Charlotte Wahl,** Noncommutative Maslov index and eta-forms, 2007
886 **Robert M. Guralnick and John Shareshian,** Symmetric and alternating groups as monodromy groups of Riemann surfaces I: Generic covers and covers with many branch points, 2007
885 **Jae Choon Cha,** The structure of the rational concordance group of knots, 2007
884 **Dan Haran, Moshe Jarden, and Florian Pop,** Projective group structures as absolute Galois structures with block approximation, 2007
883 **Apostolos Beligiannis and Idun Reiten,** Homological and homotopical aspects of torsion theories, 2007
882 **Lars Inge Hedberg and Yuri Netrusov,** An axiomatic approach to function spaces, spectral synthesis and Luzin approximation, 2007
881 **Tao Mei,** Operator valued Hardy spaces, 2007
880 **Bruce C. Berndt, Geumlan Choi, Youn-Seo Choi, Heekyoung Hahn, Boon Pin Yeap, Ae Ja Yee, Hamza Yesilyurt, and Jinhee Yi,** Ramanujan's forty identities for Rogers-Ramanujan functions, 2007
879 **O. García-Prada, P. B. Gothen, and V. Muñoz,** Betti numbers of the moduli space of rank 3 parabolic Higgs bundles, 2007
878 **Alessandra Celletti and Luigi Chierchia,** KAM stability and celestial mechanics, 2007
877 **María J. Carro, José A. Raposo, and Javier Soria,** Recent developments in the theory of Lorentz spaces and weighted inequalities, 2007
876 **Gabriel Debs and Jean Saint Raymond,** Borel liftings of Borel sets: Some decidable and undecidable statements, 2007

For a complete list of titles in this series, visit the
AMS Bookstore at **www.ams.org/bookstore/**.